T0155567

SpringerBriefs in Finance

More information about this series at http://www.springer.com/series/10282

Jan De Spiegeleer · Ine Marquet
Wim Schoutens

The Risk Management of Contingent Convertible (CoCo) Bonds

Jan De Spiegeleer
Department of Mathematics
University of Leuven
Leuven, Belgium

Wim Schoutens
Department of Mathematics
University of Leuven
Leuven, Belgium

Ine Marquet
Sint-Truiden, Belgium

ISSN 2193-1720 ISSN 2193-1739 (electronic)
SpringerBriefs in Finance
ISBN 978-3-030-01823-8 ISBN 978-3-030-01824-5 (eBook)
https://doi.org/10.1007/978-3-030-01824-5

Library of Congress Control Number: 2018958496

This Springer imprint is published by the registered company Springer Nature Switzerland AG
The registered company address is: Gewerbestrasse 11, 6330 Cham, Switzerland

Preface

The financial crisis of 2007–2008 triggered an avalanche of financial worries for financial institutions worldwide. Governments intervened and bailed out banks using taxpayers' money. Preventing such bailouts in the future and designing a more stable banking sector, in general, requires both higher capital levels and regulatory capital of a higher quality. In the new banking regulations, created in the aftermath of the crisis, the financial instruments called contingent convertible (CoCo) bonds play an important role.

The CoCo market was launched in December 2009 by the exchange of old-style hybrids into new CoCo bonds by Lloyds Banking Group. In 2010, Rabobank followed with an issue size of €125 bn. This issue was twice oversubscribed. The CoCo market experienced an exponential growth in 2013. Currently, the outstanding amount in European CoCos is above €140 bn.

CoCos are hybrid financial instruments that convert into equity or suffer a write-down of the face value upon the appearance of a trigger event. The loss-absorbing mechanism is automatically enforced either via the breaching of a particular accounting ratio, typically in terms of the Common Equity Tier 1 (CET1) ratio, or via a regulator forcing to trigger the bond. CoCos are non-standardised instruments with different loss absorption and trigger mechanisms and might also contain additional features such as the cancellation of the coupon payments.

We provide the reader an overview of the risk components of a CoCo bond and created more insights into the instruments' sensitivities. Different pricing models provided valuable information on the CoCo bond. In this book, three market-implied models are derived in detail. These models use market data such as share prices, CDS levels and implied volatility in order to calculate the theoretical price of a CoCo bond.

The sensitivity analysis of the theoretical CoCo price resulted in estimates for the sensitivity parameters with respect to the underlying stock price, the interest rate and the credit spread. These sensitivities, called the Greeks, provide the investor with insides to hedge from adverse changes in the market conditions. A performance study of the model CoCo price derived with the Greeks compared with a simple regression model indicates the importance of the credit risk in non-stress situations and the equity risk in a stress situation.

The pricing models for CoCo bonds are introduced in a market-implied Black–Scholes stock price context. Clearly, this has a drawback of assuming a constant volatility. A more advanced setting indicates the impact of this assumption. In the Heston model, a more realistic stochastic volatility context, the skew in the implied volatility surface resulted in a significant impact on the CoCo price. Hence, stochastic volatility models which incorporate smile and skew, like the Heston model, are appropriate in the context of pricing CoCos.

Furthermore, to some extend CoCo bonds can also be seen as derivative instruments with as underlying some capital ratio (CET1). In this perspective, a CoCo market price is the price of a derivative and hence contains forward-looking information or at least the market's anticipated view on the financial health of the institution and the level of the relevant trigger. This setting creates insights into the distance to trigger and enables us to determine the implied CET1 level corresponding to a coupon cancellation.

In the last chapter, a sophisticated data mining technique is applied for early-stage detection of potential risks regarding the stability of institutions by making use of market information of their issued CoCos. This method detects outliers in the CoCo market taking multiple variables into account such as the CoCo market return and the underlying equity return. Based on a robust distance in a multiple dimensional setting, we can detect CoCos that are outlying compared to previous time periods while taking into account extreme moves of the market situation as well. These outliers might require extra hedging or can be seen as trading opportunities. They could as well give regulators an early warning and signal for potential trouble ahead.

Leuven, Belgium — Jan De Spiegeleer
Sint-Truiden, Belgium — Ine Marquet
Leuven, Belgium — Wim Schoutens

Contents

1 A Primer on Contingent Convertible (CoCo) Bonds 1
1.1 What is a CoCo? 1
1.1.1 Write-Down CoCos 2
1.1.2 Conversion CoCos 2
1.1.3 Contingent Conversion Convertible Bonds (CoCoCo) 4
1.2 The Trigger Mechanism 4
1.3 Overview of the Risks 6
1.3.1 Complexity and Non-standardisation 7
1.3.2 Distance to Trigger 7
1.3.3 Non-cumulative Coupon Cancellation 7
1.3.4 Extension Risk 8
1.3.5 Recovery Rate 9
1.3.6 Liquidity Risk 9
1.3.7 Negative Convexity 10
1.4 Basel III Guidelines and CRD IV Regulation 11
1.5 Effectiveness of Issuing CoCos 14
1.5.1 Automatic Loss Absorption 14
1.5.2 Create Right Incentives 16
1.5.3 Tax Benefit 17
1.5.4 Proofs of Effect 17
1.6 Type of Investors 17
1.7 CoCo Market 18
1.8 Conclusion 20

2 Pricing Models for CoCos 23
2.1 Credit Derivatives Approach 24
2.1.1 Credit Triangle 25
2.1.2 CoCo Pricing 25
2.1.3 Recovery Rate 26
2.1.4 Probability of Triggering 27

2.2 Equity Derivatives Approach 28
2.3 Implied CET1 Volatility Model 31
2.4 Conclusion 33

3 Sensitivity Analysis of CoCos 35
3.1 Hedging CoCos 36
3.2 Sensitivity Parameters 37
3.2.1 The Greeks 37
3.2.2 Estimating the Greeks of a CoCo 38
3.3 Beta Coefficient 41
3.4 Goodness-of-Fit 42
3.5 Conclusion 49

4 Impact of Skewness on the Price of a CoCo 51
4.1 Heston Model 52
4.1.1 Pricing of Vanilla Options 53
4.1.2 Pricing of Exotic Options 54
4.1.3 Calibration 55
4.2 Case Study - Barclays 56
4.3 Sensitivity to Parameters of the Heston Model 61
4.3.1 Example of Barclays' CoCo 62
4.3.2 Distressed Versus Non-distressed Situation 63
4.4 Implied Volatility Surface 66
4.5 Conclusions 68

5 Distance to Trigger 69
5.1 Distance to Trigger Versus CoCo Spread 70
5.2 Adjusted Distance to Trigger 72
5.3 Coupon Cancellation Risk 74
5.4 Conclusion 78

6 Outlier Detection of CoCos 81
6.1 Value-at-Risk Equivalent Volatility (VEV) 82
6.1.1 Common Pitfalls 85
6.1.2 Case Study: Risk of Different Asset Classes 88
6.2 Are CoCos Moving Out of Sync? 90
6.2.1 Minimum Covariance Determinant (MCD) 92
6.2.2 Measuring the Outliers 94
6.3 Conclusion 97

7 Conclusion 99

References 103

Chapter 1
A Primer on Contingent Convertible (CoCo) Bonds

The central theme of this book is one financial instrument called a contingent convertible bond or CoCo. CoCo bonds are issued by financial institutions such as banks and (re-)insurance companies. Due to their loss-absorption mechanism, they play an important role in the new regulation guidelines after the financial crisis of 2007–2008. A CoCo bond contains an automatically loss absorption mechanism in times of crisis. This can avoid the use of taxpayers' money to save a falling financial institution in a crisis.

In this chapter an overview is given to understand the construction and financial background of CoCo bonds. First, the anatomy of the different CoCo bonds and their operating rules is explained. No standard structure has been established yet despite the issuance of CoCos from 38 different banks within European countries with a total amount outstanding closely to €160 bn by mid 2018. This underlines the importance of a detailed analysis of each new CoCo issue. The chapter contains a description of its structure, possible triggers, conversion types and the general loss absorption mechanisms. Next the current outstanding CoCo market is investigated together with the reason for their existence in the financial market and the type of investors. A research study is provided regarding the effectiveness of their loss absorption mechanism. References are De Spiegeleer et al. (2014), Maes and Schoutens (2012) and De Spiegeleer et al. (2012).

1.1 What is a CoCo?

A contingent convertible bond, also known as a CoCo bond, is a special hybrid bond issued by a financial institution. In first place, the instrument is identical to a standard corporate bond. This means that the investor receives a frequent payment of fixed coupons and will receive his initial investment back at maturity. However, when the issuing financial institution gets into a life-threatening situation, the CoCo will be written-down or convert to shares depending on the type of CoCo. The mechanism that causes the conversion or write-down is called the trigger. The trigger will as

J. De Spiegeleer et al., *The Risk Management of Contingent Convertible (CoCo) Bonds*, SpringerBriefs in Finance, https://doi.org/10.1007/978-3-030-01824-5_1

such automatically make the investor in CoCos bear part of the losses of the financial institution in stress events.

The payout of CoCos is bounded by the stream of coupon payments and the payback of the face value at maturity. This maximum payout is referred to as the bond ceiling. On the other side, the write-down of the CoCo or the conversion can lead to huge losses for the CoCo investor. Most of the time the coupon rate is a fixed level depending heavily on the healthiness of the issuing institution and typically within the range from 5 to 10% of the face value or notional amount. This relatively high rate compensates the risks of the CoCo investor. For a CoCo with a (issuer) call option, the issuer has the right but is not obliged to call back the bond at certain predefined call dates, typically at least 5 years after issuance. At a call date the issuer has the option to payout the investor the market value of the CoCo in order to cancel any future obligations of the contract. After the first call date, when the CoCo is not called, most of the CoCos turn into a floating-rate instrument. The coupon rate will from that point onwards depend on market fluctuations. More detailed information can be found in Sect. 1.3 about the risks of a CoCo.

1.1.1 Write-Down CoCos

When a (partial) write-down CoCo is triggered, the face value of the bond is written down by a predetermined fraction. The investors' wealth is now suffering a set-back. Part of the future coupons and final redemption will be lost. There is no standardised approach in this mechanism. The terms and conditions specified in the prospectus are different from country to country and from issuer to issuer. In some cases the write-down is limited to a predetermined fraction of the face value, in other cases the bond holders are completely wiped out. In January 2012, Zuercher Kantonalbank (ZKB) issued a staggered write-down CoCo. The investor could apply haircuts in multiples of 25% until the breach on the capital trigger was solved. Some contingent convertibles have a temporary write-down. Here the face value of the bond can be restored when the issuing financial institutions' health has turned positive again driven by positive financial results and adequate capital ratios.

1.1.2 Conversion CoCos

In case a conversion CoCo is triggered, the instrument will convert to a predetermined number of shares. The bond holder is forced to accept delivery of shares. The total number of outstanding shares of the institution will increase in case of an equity conversion. As a result, the existing shareholders will have a smaller, diluted part of the total outstanding equity. Hence the existing shareholders will also suffer from a conversion of these CoCos. Therefore a high dilution mechanism can create a better incentive for the risk management of a financial institution (Hilscher and Raviv 2014).

The conversion ratio (C_r) denotes the number of shares that the investor receives after a conversion. Its value is defined in the prospectus. From the conversion ratio, one can determine the embedded purchase price for each share referred to as the conversion price (C_p) of the CoCo bond:

$$C_p = \frac{N}{C_r} \tag{1.1}$$

with N the notional amount. There are various ways to specify the level of the conversion price. This choice has a significant impact on the dilution effect for the current shareholders. A decrease in the conversion price leads eventually to more shares created upon the conversion. In practice we have seen different specifications (see De Spiegeleer and Schoutens 2011 or De Spiegeleer et al. 2012). Typical settings are:

$C_p = \alpha S_0$ — The conversion price is fixed to a fraction (α) of the stock price on the issue day. Naturally one can expect that the stock price on the issue day is rather high in comparison to the stock price at the trigger. The first CoCo on the market, issued by Lloyds in 2009, had a fixed conversion price equal to its stock price on the issue day, i.e. $\alpha = 1$. Other common values are $\alpha = 2/3$ applied by Barclays Capital.

$C_p = \max(S^*, S_F)$ — where S^* is the stock price on the trigger moment, and S_F is a predefined floor price specified in the contract. By imposing a floor for the conversion price, the dilution of the current shareholders is limited. Credit Suisse implemented such a floating conversion price in their first CoCo issued in February 2011.

Notice that the recovery ratio for the CoCo investor increases in case the conversion price is closer to the share price at the time of trigger (S^*). In case the conversion price is set equal to S^*, the total value of the shares received after the trigger ($C_r S^*$) is equal to the notional. This leads to a 100% recovery for the CoCo investor.

These settings show that it is extremely important to set the conversion price above the share price at the time of trigger in order to let the CoCo holder absorb part of the losses. In the opposite case the initial shareholder would suffer from the high dilution effect and the CoCo holder would not absorb losses except the loss of all future coupon payments. A conversion price equal to the stock price on the trigger moment is not allowed in practice, i.e. $C_p = S^*$. This floating conversion price equals the share price at the time of a financial distress for the issuing bank and is as such assumed to be rather low. In fact the CoCo holder does not absorb any losses since the total value of the shares received equals the notional value. For this reason this type of conversion price is not allowed in practice.

1.1.3 Contingent Conversion Convertible Bonds (CoCoCo)

Standard Contingent Convertible bonds consist of a corporate bond as host instrument. Without a trigger event, the CoCo holder receives coupons on a frequent basis and the notional amount at maturity. One exception has been issued against the standard coupon bearing debt host instrument. This different type of CoCo was issued by the Bank of Cyprus in February 2011 under the name Convertible Enhanced Capital Securities (CECS). The host instrument of this product was a convertible bond and it received the term contingent conversion convertible bond (or in short CoCoCo) by the financial industry (Campolongo et al. 2017).

The holder of a convertible bond has the right to convert his initial investment to a predetermined number of shares. The investor in such a type of bond will only convert if the shares have increased a lot in value (and are assumed to stay so). Notice the main difference with CoCos where the conversion is forced by the issuer or the regulator in bad financial situation for the issuing company.

A CoCoCo is a combination of both worlds. It gives the investor the opportunity to participate also from the price appreciation which is not possible with a CoCo. Like every convertible bond, the investor can decide at predefined times to convert the bond to equity and gain from profits in the underlying shares. But like a CoCo, a mandatory conversion can also appear in bad times. The investor has to absorb (part of) the losses. Due to the extra possibility to participate in the upside for the investor of a CoCoCo, the coupon rate can be fixed at a lower level compared with a 'standard' CoCo. An in-depth analysis into the valuation and the dynamics of this innovative financial market product is given in Campolongo et al. (2017).

1.2 The Trigger Mechanism

The trigger of a CoCo is documented at length in the prospectus. It defines when the bond will get converted or is written down. The existence of such a life-threatening situation is typically measured in terms of an accounting or capital ratio falling below a pre-defined level. Hence the capital ratio corresponds with a measure of the healthiness of the bank's balance sheet.

The account ratio reflects the amount of regulatory capital compared to the risk-weighted assets (RWA) on the balance sheet. The RWA is the total sum of the assets multiplied with their corresponding risk weight. The weighting scheme is used to take the risks for each asset class into account. In order to manage the risks, banks are obliged to hold enough regulatory capital to protect against a decrease in value of the assets.

An example of an accounting ratio typically used in Basel III, and also for CoCos, is the bank's Common Equity Tier 1 (CET1) ratio. The CET1 ratio is defined as a measure of a bank's common equity capital expressed as a percentage of risk-weighted assets. The Basel Committee proposes an absolute minimum level of this ratio at 4.5%. The trigger levels for the CoCos based on their CET1 ratio range in most cases from 5 to 8%.

Since the risk weighting scheme takes place under supervision of the national regulator, the exact definition of the CET1 ratio varies across different domiciles (Hajiloizou et al. 2014, 2015). However the Basel Committee of the Bank of International Settlement (BIS) helps to create a consistent approach across different domiciles. Nevertheless one must be cautious with interpreting the levels of the CET1 ratio. In Merrouche and Mariathasan (2014), the authors document that weakly capitalised banks are likely to manipulate risk weights more severely. This could lead to manipulation in the triggering as financial institutions can be creative with their definition of capital ratios. For example Dexia bank reported a Tier 1 ratio above 10% before being rescued by the government in October 2010 and also Lehman Brothers reported a CET1 ratio of 11% before it went bankrupt.

Under Pillar 3 of the Basel II framework, large banks are subject to minimum disclosure requirements with respect to defined key capital ratios on a quarterly basis, regardless of the frequency of publication of their financial statements (Basel Committee on Banking Supervision 2014). This leads to a next drawback of this type of trigger. A potential danger is that this trigger may be activated too late due to the fact that the accounting ratio is not continuously observable (Flannery 2009). However, this does not mean that the loss absorption mechanisms such as a write-down or conversion into shares can only be activated on a quarterly basis. CoCos can be triggered at any point in time, banks can (be forced to) disclose their capital levels at any time.

Moreover, most CoCo bonds have also a regulatory trigger controlled by the bank's national supervisor. The national authority has the discretion whether or not to trigger the bond (Basel Committee on Banking Supervision 2010c). A national regulator will trigger a CoCo because action is necessary to prevent the bank's insolvency. Therefore the point of triggering by the regulator is often referred to as the point of non-viability (PONV). This regulatory trigger incorporates an extra difficulty in the instruments since the point of trigger is now even harder to access and could reduce the marketability of the instrument. Opponents state that this regulator trigger mechanism is a blank cheque written out to the financial authorities since the trigger might occur based on non-public, supervisory information.

Besides the above discussed trigger mechanisms present in the currently issued CoCo market, also other possible mechanisms have been discussed in the literature. For example, market based triggers were proposed in Flannery (2009), Hilscher and Raviv (2014) and Calomiris and Herring (2011). A market trigger is based on an observable issuer-related metric reflecting the solvency of the issuer such as the stock price or credit default swap (CDS) spread. It is a more transparent way to define the trigger as its value is continuously observable and direct hedge instruments are available. The market based trigger is also more forward looking whereas an accounting ratio is typically backward looking (Haldane 2011). On the contrary, extreme events on the stock market (like a flash crash) can create a market based trigger without changing the capital structure of the issuing financial institution. Also stock price manipulations and hedging of the CoCo bond could result in a trigger (see Sect. 1.3.7). Indeed in D' Souza et al. (2009), the authors denote that market based triggers might actually aggravate bank runs, rather than prevent them. These disadvantages explain why no CoCo has been issued so far with a market based trigger.

Various other trigger mechanism can be found for CoCos in the literature. For example, in Madan and Schoutens (2011) an alternative trigger is introduced based on capital shortfall. In Calomiris and Herring (2013) the health of a systemically important financial institution (SIFI) bank is expressed by the quasi-market-value-of-equity ratio. This new ratio is a 90-day moving average of the ratio of the market value of equity to the sum of the market value of equity and the face value of debt. Other examples create multiple trigger mechanism to avoid certain pitfalls of the construction of a CoCo (De Spiegeleer and Schoutens 2012b). This concept of a dual trigger is derived from the mindset that CoCo triggers should be contingent on both individual bank and systemic measures. A combination of multiple criteria that need to be satisfied for a CoCo to trigger are described in McDonald (2013) and Allen and Tang (2016).

1.3 Overview of the Risks

CoCo bonds can be seen as strategic funding tools. They provide the issuer with an extra regulatory capital buffer to prevent systemic collapse of (other) important financial institutions. They increase the safety and soundness of the global financial system and provide better incentives for the management of the financial institution. A CoCo bond contains an automatically loss absorption mechanism in times of crisis. This extra capital buffer can avoid the use of taxpayers' money to save a falling financial institution in a crisis.

Hence CoCos are designed to fail, without bringing down the bank itself in the process. The loss absorbing character of the CoCo bonds creates a risk, that on a trigger the owner will suffer a full or partial loss on the face value. This explains why these bonds have a high coupon. The trigger activates the loss absorption mechanism and the CoCo investors will bear this loss. The CoCo can absorb losses in two ways either by conversion into equity (worth less than the bond's notional) or by imposing a (full or partial) write-down of the face value.

The risks included in Contingent Convertible bonds are often a point of discussion. Next to the risk of triggering the loss absorption mechanism for the CoCo holder, these instruments contain extra and more hidden features causing risks. The multiple risk factors in the CoCo bond are discussed in this section.

1.3.1 Complexity and Non-standardisation

The risk profile of a CoCo corresponds to an investment product with a low probability of a high loss and a high probability for a moderate gain. The moderate gain consists out of receiving a high coupon and the face value in case of no trigger. From a risk perspective, one could say CoCos have a limited up-side, given by the bond feature, and a full down-side potential.

In the UK a temporary product intervention rule has been put at work on these types of bonds for retail investors excluding professional, institutional and sophisticated or high net worth retail investors. The Financial Conduct Authority (FCA) of the UK states that CoCo bonds are too complex, include unusual loss absorption and contain high risks for the investors. Their worries go out to the individual investor who will blindly follow the high yield, especially in a current low interest rate environment. This restriction on sales and marketing became into effect on October 1, 2014 (FCA 2014). In October 2015 this regulation was replaced by a permanent restriction to sell or promote (or approve promotions) of CoCo bonds to ordinary retail clients excluding high net worth investors or sophisticated investors along with an institutional investor (FCA 2015). High notional amounts are in general already applied to CoCos in order to protect small retail investors from entering the CoCo market without understanding the risks.

1.3.2 Distance to Trigger

Although the definition on the point of triggering was given in previous section, some difficulties arise with this time of conversion or write-down. The exact definition of the CET1 ratio can differ between different issuers and the CET1 ratios are only made available on a quarterly basis. While trigger factors are contractually defined, there remains uncertainty due to the enforced write-down or conversion by the regulation authority in case of non-viability or resolution actions. Furthermore, most pricing models for CoCos take the accounting trigger into account but have no idea of the time when the regulator will trigger these bonds. This lack of knowledge is often translated to ignoring the point of non-viability. This can induce a high risk in modeling the trigger event. An investigation of this model risk is given in Chap. 5.

1.3.3 Non-cumulative Coupon Cancellation

CoCo bonds are loss absorbing by construction and this particular feature allows them to count as regulatory capital. In a Basel III setting, these bonds can count as Tier 2 (T2) or as Additional Tier 1 (AT1) bonds. AT1 CoCo bonds have a more permanent character given the fact they are perpetuals. The first call date has to be at least 5 years after the issue date of the bond. A particular property of the coupons distributed by such an AT1 CoCo bond is the fact that these coupons might be cancelled. Such a cancellation would not be considered as a default, in contrast with the cancellation of coupon payments on T2 bonds or senior bonds. Furthermore, there is no incentive (besides reputation) included for the issuer to pay coupons.

Also the pay-out of coupons of the CoCo should not be related with the dividend payments. In the EU it is not allowed to include dividend pushers or dividend stoppers in the contract of an AT1 CoCo. A dividend pusher would force the coupon payment on the bond in case dividends on the shares are paid out. A dividend stopper is the opposite mechanism where the dividend payment is prohibited if the coupon payment is cancelled. Such mechanisms forcing a relation between dividend payments and coupon payments of the CoCo are not allowed.

The coupons of an AT1 CoCo bond are non-cumulative in a sense that the cancelled coupon payment is lost forever. Since coupon rates are high, this would cause a major impact for the investor. On the other side, no CoCo coupon payment has ever been cancelled in the history of CoCos. Notice that this non-cumulative procedure is in contrast with the dividend and bonus payments, which can be reimbursed with higher payments when the institution returns to health. In the first quarter of 2016 concerns emerged that Deutsche Bank would have a lack on available cash based on the maximum distributable amount (MDA) measures and might block coupon payments of its CoCo bond. This created a turbulent financial market in the beginning of 2016 while Deutsche Bank had to reassure its coupon payments of its outstanding CoCos. This occurrence in the financial market created a proposal by the European Commission in Brussels to prioritize the AT1 coupons over other discretionary payments. However the European Banking Authority (EBA) objects the prioritizing of CoCo payouts (Weber et al. 2017).

1.3.4 Extension Risk

A risk included, specifically for the AT1 CoCos, is that the issuer might not buy back the CoCo as soon as expected by the investor. CoCo bonds containing one or more call dates have an unknown maturity. The redemption of the bond is controlled by the issuer who might for example decide to extend the bond on a call date. A callable Contingent Convertible bond has typically a split in its coupon structure: a fixed coupon distributed before the first call date and a floating coupon after this date.

Incentives to redeem are not allowed as stated by the Basel III requirements. The fact that there are no step-ups on the floating coupons, increases the probability of an extension. The presence of such coupon step-up would indeed be seen as an incentive to redeem the bond prematurely and would weaken the permanent character of the AT1 contingent convertible (De Spiegeleer and Schoutens 2014). Hence there is no longer the presence of an important step-up penalising the issuer when extending the bond to the next call date.

1.3.5 *Recovery Rate*

The recovery of a conversion CoCo depends on the conversion type together with the conversion price and is often a point of discussion. For example the recovery value of a conversion CoCo investor depends on the share price. The share price can drop further after a conversion and can eventually become worthless. However if the bank survives and the share price increases, the investor, now a stockholder, benefits from the recovery value upside. Also step-up write-down CoCs can recover from a financial distress.

Up to date only one CoCo has ever been triggered for which we can observe the true recovery rate. In June 2017 the authority forced a resolution for the Spanish bank Banco Popular. Together with the resolution program, the CoCo of Banco Popular was triggered. Although this CoCo had a conversion to shares, the recovery was zero since the share price became worthless.

1.3.6 *Liquidity Risk*

Liquidity relates to how fast one can buy or sell a product on the financial markets at a stable price. It represents the trading activity of the instrument. CoCos are a relatively new asset class with only limited liquidity and still a lot of regulatory and model uncertainty. In Allen (2012), the author expresses her concerns regarding a nearly triggering of the CoCo. As an effect the market liquidity might decrease due to panic selling. Furthermore larger negative swings might be caused on equity due to the low liquidity of debt compared with equity.

The bid-ask spread of the CoCos can provide a first inside in the liquidity. This spread captures the difference between the highest price a buyer is willing to pay (bid) for a bond and the lowest price that a seller is willing to accept (ask). Hence a lower bid-ask spread indicates a higher liquidity. For example in February 2016 the bid-ask spread of Deutsch Bank 6% CoCo bond tripled from 0.6% in 2015 up to 1.5% due to the market turmoil (Mehta 2016). In general the CoCo market was less liquid around the end of 2011, June 2013 and in the first quarter of 2016 (Fig. 1.1). The higher spread in these time periods suggests a more illiquid market.

Multiple features of the liquidity of CoCo bonds are discussed in Hendrickx (2016–2017). First, CoCos with a larger amount outstanding tend to be more liquid. Indeed when more CoCo bonds are issued, more and bigger trades can be realized. Second, the liquidity increases as the next call date comes closer. Third, in case the asset swap spread is high the CoCo seems to be less liquid. This can be related to the fact that a high asset swap spread indicates a higher risk and hence less investors might be interested in the product.

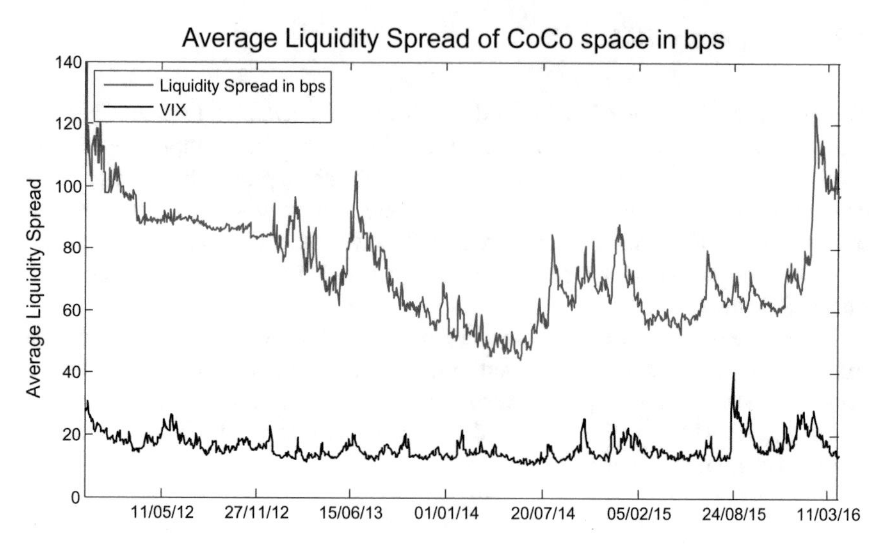

Fig. 1.1 Bid-ask spread of the CoCo market versus VIX

1.3.7 Negative Convexity

If the stock price moves down, one naturally can expect to see as well a negative impact on the value of a CoCo, in particular when the stock price is under severe stress. If the stock price falls, the financial institutions are more likely to enter a more difficult financial situation. The rate of change of a CoCos' model price corresponding to the underlying share is called delta (see Chap. 2):

$$\Delta = \frac{\partial P}{\partial S} > 0 \tag{1.2}$$

Due to the positive relation between the CoCo model price and the underlying stock, the delta of a CoCo is always positive. A more sophisticated investor could proceed to hedge a holding in CoCos based on this information. These investors will sell short shares of the bank. As such the potential loss in the CoCo bond can (partially) be offset by the gain in the short position of the stock. A short position in the stock can protect the investor against the loss when the share price falls (De Spiegeleer and Schoutens 2012a).

A small change in the stock is much more crucial for a CoCo investor if the stock price is already low since there is a higher chance that this small change will result in a trigger event. Hence, the equity sensitivity increases when the stock price S drops. We can express this by the second order sensitivity, denoted by gamma:

$$\Gamma = \frac{\partial^2 P}{\partial S^2} < 0 \tag{1.3}$$

As the shares drop, the more sophisticated investor will start selling more shares to hedge his position. But selling shares in a falling market is going to pressure shares further. This way the investor is pulled into the uncomfortable situation where he is forced to sell more as the shares drops, leading on his turn to extra drops in the share price in case of a poor liquidity. This is called the downward spiral or death spiral effect.

An issuer has to pay attention to the issue size in order to make hedging possible for the investors without ending in this share price collapse. The free float of shares outstanding and the average daily traded volume of shares are therefore important bottlenecks on the CoCo issuance. A solution for the spiral effect can be to include multiple triggers (De Spiegeleer and Schoutens 2012b, 2013) and (De Spiegeleer et al. 2014). Each time a trigger is hit, a part of the product will be converted. This mechanism has not been issued so far. Another solution is the coupon cancellation feature as explained in Corcuera et al. (2014).

1.4 Basel III Guidelines and CRD IV Regulation

The Basel Committee, founded in 1975, formulates general supervisory principles and proposals for financial institutions. It attempts to set an international regulation by exchanging information on national supervisory arrangements. The principles of the Basel Committee can be used as a basis in the requirements on a national level but are not obligatory (Basel Committee on Banking Supervision 2013). An example of the interpretation of the Basel requirements in the European Union is the Capital Requirements Directive (CRD). These directives are mandatory for countries of the EU and need to be implemented into the national law.

The first Basel Accord (Basel I) from 1988 was the first attempt to set international regulation and took only one type of risk into account, namely credit risk. The assets of financial institutions were categorized into five risk categories, i.e. 0, 10, 20, 50 and 100% which denoted their risk weight. For example, governments bonds were considered to be harmless and received a zero weight, while loans to corporations received the highest weight. Under Basel I, banks were required to keep a capital level equal to at least 8% of their risk weighted assets (RWA). This ratio of capital and RWA is also called the Cooke ratio. In Basel I the capital was divided into two components called Tier 1 capital and Tier 2 capital. Tier 1 capital is used in a going concern context. Whereas Tier 2 capital or supplementary capital will be used in a gone concern to prevent default (Hull 2010).

In 2004 the Basel Committee published new requirements to deal with different types of risks like operational, market and credit risk. These requirements of Basel II were classified in three different pillars: minimal capital requirements with adapted weighting schemes, supervisory review and market discipline. Pillar 3 is based on the idea that market participants, if they have more information, will push the banks to deal with better risk management.

During the financial crisis, it became clear that there were still shortcomings in Basel II. This has resulted in new and more restrictive requirements in the third Basel Accord issued in December 2010. Basel III focuses on the quality of the capital together with liquidity of a bank. The guidelines of Basel III are translated in 2013 to European directives, summarised under the name CRD IV, which will be fully implemented by 2019.

Avoiding the need for and cost of bail-outs with taxpayers' money in the future, demands both higher capital levels and better loss absorbent capital. The high quality loss absorbers are called Common Equity Tier 1 (CET1) and consist of common equity and retained earnings. The preferred stock and perpetuals became Additional Tier 1. The Additional Tier 1 capital should help a bank to remain in a going concern and is tangeable to common equity. Examples of Tier 2 capital are hybrid instruments and subordinated debt (Campolongo et al. 2017).

The idea of dividing the capital into more categories, emerged from the penalties in hybrid securities when it came to loss absorption in a going-concern basis as effectively as previously expected. These hybrid securities failed their function as the management did not want to disappoint the investors. This is not possible with a CoCo since the coupons are automatically reduced in times of stress (Basel Committee on Banking Supervision 2010a). In a proposal of August 2010, the Basel Committee imposed the fact that debt instruments can only count toward regulatory capital when they can absorb losses in a state of non-viability (Basel Committee on Banking Supervision 2010c). In January 2011 this was even further restricted to debt instruments that absorb losses such that no taxpayers' money will be needed to bail out the bank.

The minimum percentages of RWA from Basel III for the different capital classes are shown in Fig. 1.2. In Basel III, the CET1 capital is required to be at the level of at least 4.5% of RWA. Extra buffers, called combined buffers requirement are entered in the CRD IV regulation.[1] This requirement consist of a capital conservation buffers[2] and countercyclical buffers[3] and bring the minimum capital requirement for CET1 up to 9.5% of RWA under Basel III. The combined buffer requirement includes also extra systemic risk buffers[4] and/or the systemically important institution buffers.[5] For globally systematic important banks (G-SIBs), this value can be increased up to 13% of RWA by an extra buffer. Also domestic systemically important banks (D-SIBs) identified as systemically important bank by a national regulator, can be forced to hold extra capital buffers.

However, CoCo bonds are not allowed to serve as part of the combined buffer requirements. In the new regulations of Basel III, CoCo bonds can only be part of Additional Tier 1 or Tier 2 bonds (Corcuera et al. 2013). Under CRD IV Additional Tier 1 (AT1) CoCos can account for 1.5% of RWA, while Tier 2 CoCos can account

[1] According to Art 128 (6) CRD.

[2] According to Art 129 CRD.

[3] According to Art 130, 135-140 CRD.

[4] According to Art 133-134 CRD.

[5] According to Art 131-132 CRD.

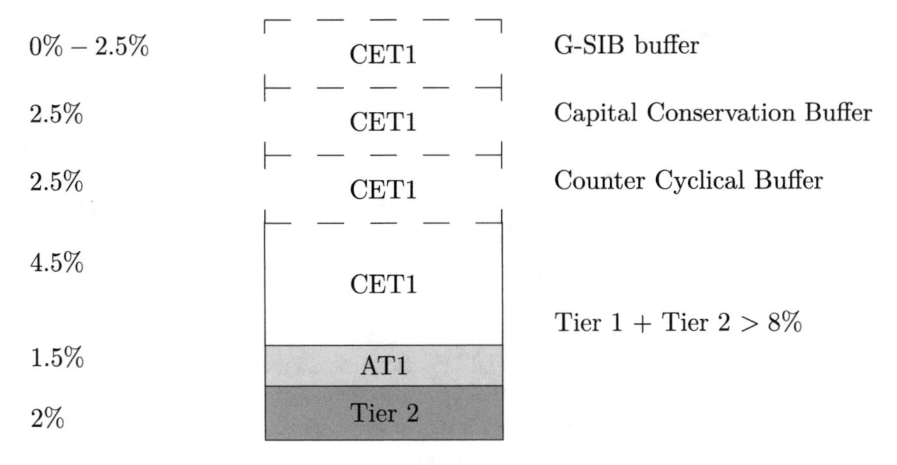

Fig. 1.2 Basel III: minimum percentage of RWA

for 2% of the RWA (Maes and Schoutens 2012). A lot of restrictions are given for a CoCo bond to qualify as an AT1 bond. For Additional Tier 1, the CET1 trigger needs to be at least equal to 5.125% and also a regulatory trigger needs be contained in the CoCo. The product has to be perpetual in maturity with no incentives to redeem early and coupon payments can be suspended (Avdjiev et al. 2013). The first CRD IV compliant loss absorbing bond with a qualification as Additional Tier 1 was a conversion CoCo of BBVA, a Spanish bank, with multiple accounting triggers. The conversion price has a floor and the maturity of this CoCo is perpetual but becomes callable after 5 years.

Furthermore the European Banking Authority (EBA) included an extra regulation in the CRD IV that requires institutions to meet the combined capital buffer. In case the combined capital buffer is insufficient, banks are required to calculate the Maximum Distributable Amount (MDA). These banks are, before the calculation of the MDA, prohibited to make payments on AT1 instruments, including AT1 CoCo bonds. Hence, if the combined capital buffer is insufficient in combination with a low MDA, the coupon distribution might be limited or suspended.

Early 2016 this mechanism raised confusion on the market regarding the multiple approaches of calculating a banks' MDA and the uncertainty for the payments of bank debt instruments (ECB-public 2016). The MDA is calculated as a factor of the sum of interim and year-end profits. The factor equals 60%, 40%, 20% or 0% depending on which quartile of its combined buffer the firm is in.[6]

Also, the Supervisory Review and Evaluation Process (SREP), a harmonised tool of banking supervision across the euro area, has included an extra Pillar 2 Requirement that needs to be fulfilled in terms of CET1 capital.

[6]Described in Art 141(2) to 141(10) of the CRD Directive 2013/36/EU.

1.5 Effectiveness of Issuing CoCos

The financial crisis of 2007–2008 was the unfortunate outcome of a widespread disruption in the global financial world. This distress caused a real domino-effect given the interconnectedness of financial institutions across the globe. After the collapse of Lehman Brothers, governments intervened and bailed out banks with taxpayers' money in case of non-viability to protect the financial system from worse.

The bail-out of a bank occurs when the national authority steps in to save a bank with the use of taxpayers' money. With the use of taxpayers' money, governments can bail out a bank by increasing the deposit-insurance level, including guarantees for certain debt assets or direct capital injections to increase the equity buffer of a bank (Nordal and Stefano 2014). However this imposes clear problems with the market discipline of the bank. The bank might take on more risks in order to gain more profit without being concerned of a failure. Hence, this idea of too-big-to-fail (TBTF) creates a moral hazard of implicit government support. In order to strengthen the banking sector and to avoid further bail-outs of banks financed with taxpayers' money, more regulatory capital was going to be needed. On top of this, this capital had to be loss absorbing.

1.5.1 Automatic Loss Absorption

CoCos can work like a shock absorber for the banks in times of stress. They automatically improve the solvency when it would otherwise be difficult to raise capital levels. This reduces the cost of governmental bail-out and prevents us from a systemic collapse of important financial institutions. The design of CoCos stands in contrast to the penalties of the hybrid Tier 1 bonds during the financial crisis of 2007–2008. These hybrid securities failed their function as the management did not want to disappoint their investors. This is not the case for CoCos since their loss absorption mechanism is triggered automatically (Basel Committee on Banking Supervision 2010b).

The investment of banks in multiple assets is funded by the liabilities side of the balance sheet. These liabilities consist of deposits, debt instruments, equity and a combination of equity and debt instruments called hybrid security instrument. In Fig. 1.3 we illustrate the balance sheet of a bank. In case the bank experiences losses from its business and financial risks, the bank's assets and liabilities will decrease in value. First the equity holders will absorb the losses. If there is no equity and as a last solution before defaults, extra losses will be absorbed by the hybrid instruments. In case the bank can not payout the interest payments to its debt holders, it will be declared bankrupt.

Since CoCos are hybrid instruments, they reinforce the banks' balance sheet with an extra buffer included to absorb losses. When a bank gets into a non-viable state, the balance sheet can be quickly and automatically reinforced by triggering

Fig. 1.3 Balance sheet of a bank (Marked: hybrid debt)

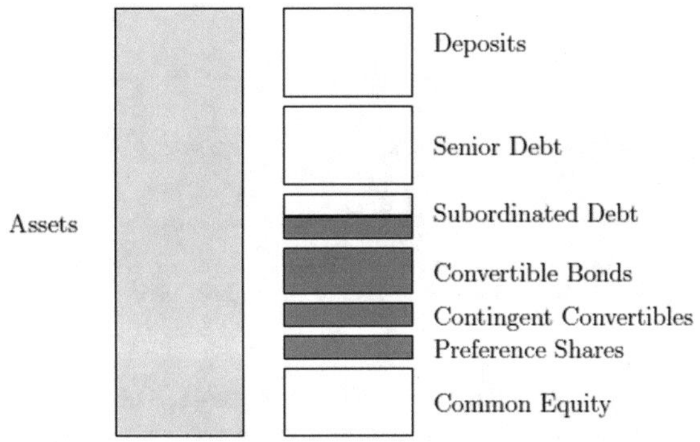

the conversion or write down of the CoCos. Moreover, the coupon stream is also cancelled. All this can reduce significantly the debt and improve the bank's liquidity. This explicit face value conversion (or write down) depreciates the standard implicit guarantee of the government bail-out.

The capital hierarchy can also be observed from the yield in each different type of instrument. The spread is the extra yield above the zero-coupon treasury yield curve in order to make the discounted present value of future cash-flows equal to its present market price. The price of a CoCo with maturity T can be expressed in terms of the CoCo spread z by:

$$P = \sum_{i} C_i e^{-(r+z)t_i} + N e^{-(r+z)T} \tag{1.4}$$

with the continuous risk free interest rate r, coupon payments C_i at time t_i and notional N. The CoCo investor will typically ask for higher yield than the yield of the senior debt issued by the same financial institution due to the higher risks (De Spiegeleer and Schoutens 2014). For example, the equally-weighted averaged AT1 spread can be compared with the high yield bond indices and a global emerging market corporates index in Fig. 1.4. In the first quarter of 2016 the yield of the CoCo market shoots above all high yield bond indices, indicating the increased concerns with the CoCo market and its relation with the underlying stock market.

Although there is a clear ranking of the CoCo bonds at first sight, discussion exist about their ranking near triggering or coupon deferral. The moral hazard of the management of a conversion CoCo is in question. By triggering a CoCo, the management will hurt CoCo holders while shareholders might still maintain their equity interest (Nordal and Stefano 2014). Furthermore, a coupon cancellation could potentially put CoCo-holders in a worse position than shareholders. In theory this suggest that under certain circumstances the yield of such CoCo might be higher than the expected returns of the underlying equity. This makes the ranking less clear than shown in Fig. 1.3.

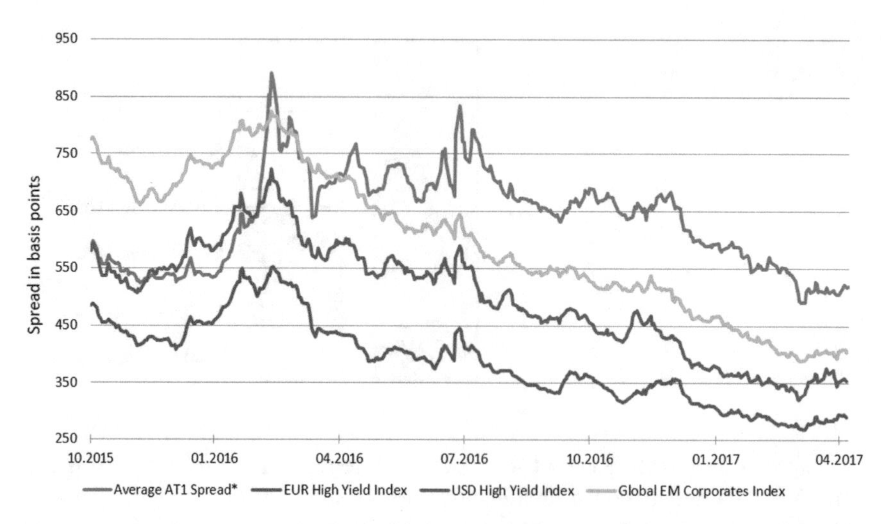

Fig. 1.4 CoCo spreads compared to high yield bonds. Source: Bloomberg, Merrill Lynch Indices, own calculations. (*) Equal weighted average of all European AT1s, which were issued before October 1, 2015

1.5.2 Create Right Incentives

CoCo bonds work in between a going or gone concern situation and can be triggered by the bank capital ratios or by a regulators decision. This is typically in contrast with the bail-in feature of certain debt instruments which is always a gone concern. Debt instruments with a bail-in feature will also be converted or written down but only in case the country's resolution authority steps to announce a bankruptcy situation. Issuing a CoCo in a going concern reduces the risk incentives. The CoCo debt is as such superior to subordinated (bail-in) debt in terms of discouraging the risk choices (Martynova and Perotti 2015).

For a conversion CoCo, the perspective of diluting the existing shareholders is a good incentive to keep the bank solvency away from the trigger. In addition, an efficient loss-absorption potentially reduces the systemic risk, sending a clear signal to the market.

Furthermore, the behaviour of risk taking could be reduced by issuing bonuses to the trading staff in terms of CoCos. As such the staff would also share in the downside and the revenues would stay in the business. By a renumeration in CoCos, the management would become more risk adverse. G Haldane stated that the capital ratio would increase by 1% if 50% of the bonuses in 2000-2006 in the UK would have been paid out in CoCos (Haldane 2011). Credit Suisse even started in 2014 to pay part of its top bankers' deferred bonuses in CoCo bond equivalents ensuring their interests lie in the bank's long-term stability.

1.5.3 Tax Benefit

The hybrid instrument might lower tax cost of capital for the issuers. Typically the more an instrument behaves like debt, the more likely it is to be tax deductible. For example the interest rates of Rabobank CoCo are tax deductible for Dutch tax purposes. In Belgium there is a requirement that the issuer must have full discretion to cancel distributions in order to achieve deductibility. This may be a significant difficulty (Fiamma et al. 2012). The tax treatment of CoCos varies by jurisdiction and is typically non uniform across national tax laws. Hence the effect of issuing the CoCo bond in terms of tax deductibility due to the debt side of the instrument is very difficult to generalise.

1.5.4 Proofs of Effect

Based on the Value-at-Risk (VaR) and Expected Shortfall (ES) estimates for the issuer's default risk, the authors in Jaworski et al. (2017) acknowledge the improvement in the solvency of the bank by issuing CoCo bonds. These bonds strengthen the resilience of the issuer under the condition that the probability of conversion triggering is higher than the significance level of VaR. Furthermore the effectiveness of issuing a CoCo bond can be related to the boosting of regulatory capital of the issuing financial institution and their automatically loss absorption mechanism.

In Benczur et al. (2016), the authors estimate the effectiveness of the new EU regulatory framework. They estimated the financial cost considering a crisis of a similar magnitude as in 2007–2008 but taking into account all loss absorbing mechanisms of a bank. This safety-net included the explicitly modelling enhanced Basel III capital rules, the bail-in tool and the resolution funds. Their research study showed that potential costs for public finances decrease from roughly 3.7% of EU GDP (before the introduction of any new tool) to 1.4% with bail-in, and finally to 0.5% when all the elements we model are in place. This latter amount is indicated to be very close to the estimate of leftover resolution funds and the size of the Deposit Guarantee Scheme.

1.6 Type of Investors

Investors in CoCos are mainly driven by the high yield in today's environment of around zero real interest rates. The high yield will serve as a compensation for the risk taken by the investor. The yield in AT1 CoCos is around 7 and 6% for T2 CoCos (Source: Bloomberg). The group of CoCo investors consists mainly of sophisticated retail investors and small private banks from Europe and Asia. The number of investments depends on regulations and credit ratings. US institutional

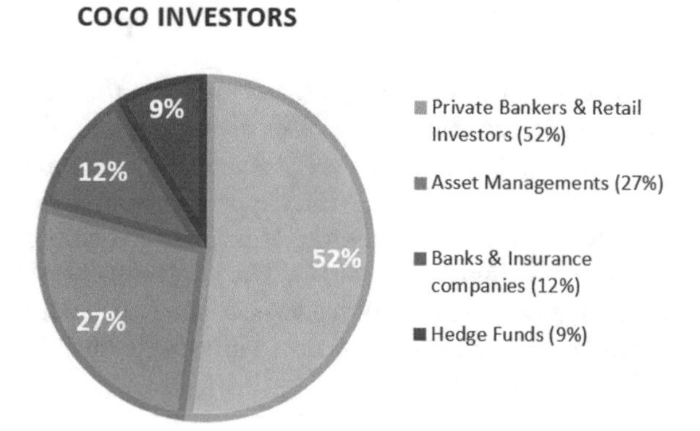

Fig. 1.5 Investors in CoCos. Source: The Handbook of Hybrid Securities (De Spiegeleer et al. 2014)

investors are another group of investors who are driven to find alternative investment classes (Avdjiev et al. 2013).

A potential danger is that CoCos do not reduce the domino effect as was observed in the last financial crisis. They could even increase the interconnectedness and the corresponding contagion risk. If CoCos are mainly held by financial institutions, no redeuction of the systemic risk is achieved. The trigger can create more triggers if CoCos are held by other banks. However by regulation, financial institutions are discouraged to invest in other CoCo bonds based on extra capital requirements (Liberadzki and Liberadzki 2016). Hence the solution could be to limit CoCos sales to other financial institutions and market it to the different investment groups like hedge funds, private banks or high net worth individuals, family office, etc. As can be seen in Fig. 1.5, this idea is put to work and the overall loss absorption of CoCos is not transferred to other financial institutions. The main part of the investors group consists of retail investors and small private banks.

Most of the issues of contingent capital were multiple times oversubscribed by so-called sophisticated investors. This type of investor has more investing experience and knowledge to weigh the risks and merits of an investment opportunity compared to standard retail investors. In general, asset managers have large existing holdings in Tier 1 and Tier 2 products. They have the expertise in-house and have been exposed to hybrid instruments before 2008. Multiple CoCo funds were created by asset managers. There is also a small interest from hedge funds which have more risk-appetite. Furthermore they have the experience to hedge to some extend the unwanted risks away. Employees of some investment banks like Barclays Capital and Credit Suisse pay CoCo bonuses to their top-management (De Spiegeleer et al. 2014).

1.7 CoCo Market

The first financial CoCos were introduced by the Lloyds Banking Group in December 2009. In November 2009 Lloyds had to raise extra capital in order to avoid entering into the UK Asset Protection Scheme consisting of the payment of fees to the UK

Treasury in change for an asset relief (Maes and Schoutens 2012). Actually it was not a typical raise of new capital but rather an exchange offer for certain existing hybrid debts. Since Lloyds had received state aid, the institution was not allowed to redeem coupons to the hybrid capital investors. Therefore their investors were offered to swap some of their hybrid instruments for CoCos (called Enhanced Capital Notes (ECN)). The issue size was equal to \$13.7 bn. The CoCos would convert into equity in case the Core Tier 1 capital of Lloyds dropped below 5%. In the new Basel III definitions, this corresponds more or less to a CET1 ratio dropping below 2.5% (De Spiegeleer and Schoutens 2010; De Spiegeleer et al. 2014).

In 2010 also Rabobank issued CoCos under the name 'capital securities' with an issue size of €1.25 bn. The issue was twice oversubscribed. Since Rabobank was not listed, the CoCos had a write-down loss absorption mechanism which in case of a trigger would wipe out 75% of its notional and the other 25% of the notional would be paid back to the investor. Hence for this specific write-down CoCo bond the contract would immediately mature after the time of trigger. The trigger was expressed in terms of the equity capital ratio which is the ratio of the equity capital to the risk weighted assets and was set at 7%.

At the end of 2010 CoCos gain more credibility due to the new set of capital requirements of the Basel Committee. Because of new regulations in Switzerland called 'Swiss Finish', Credit Suisse issued a conversion to equity CoCo in February 2011. The new requirements proposed a total capital of 19% of the risk weighted assets as by Basel II with 10% held in the form of common equity. The other 9% could be satisfied by issuing CoCos (Commission of Experts 2010). The total size of the issue of Credit Suisse was \$2 bn and had a massive over-subscription.

The asset quality review and the subsequent stress test of the European Central Bank (ECB), also allowed CoCo bonds. Because of this regulatory framework, banks have been embracing these new hybrid instruments. In September 2013 Credit Suisse issued the first CoCo bond in euro currency and opened the European CoCo market. From that point onwards a lot of other banks followed with their own issues of CoCo bonds.

Different drivers, next to increasing the capital buffers, exist for financial institutions to issue CoCos. For example UBS and Credit Suisse were driven by new regulation guidelines of their national authority introduced in 2013 and need to be fully implemented by 2019. This new regulation includes a basic requirement capital of 4.5% CET1 capital, a buffer capital of 8.5% in CET1 or high trigger CoCos with trigger level of 7% and a progressive component with up to 6% in CoCos containing a low trigger level of 5% only (Nordal and Stefano 2014). The funds of KBC from issuing CoCos in January 2013, were used to repay part of the government bail-outs that the bank received in 2008 and 2009. KBC opted for write-down CoCos to avoid a further dilution effect for the strategic shareholders. In May 2014, Deutsche Bank issued its first CoCos which were more than five times oversubscribed. Three types of CoCos were issued with a total value of €3.5 bn. Deutsche Bank has waited with CoCo bonds until it was confirmed that the coupons would be tax deductible (Thompson 2014). In the US, no CoCos have been issued so far. The Dodd-Frank act made a statement that these bonds would not contribute for the banking system in the US. Furthermore, interest rates of convertible debt is not tax deductible in current Treasury guidance.

More recently the impact of Basel III guidelines is observed by an increase in the capital buffers. As a result the CoCo market boomed in 2015 with multiple issuers entering the CoCo market. In the first quarter of 2016 stress in the stock market related to the uncertainty in coupon payments for Deutsche Bank interrupted this booming market. The first new issue of 2016 was for UBS in March. For this issue with size \$1.5 bn, there were \$8 bn of orders. Also in August 2016 the financial market became again clearly aware of the risk in the CoCo market with the relation to the underlying stock market price due to the equity side of these hybrid bonds. In June 2017 the first CoCo was triggered. The CoCo of the Spanish bank Banco Popular was triggered as part of the resolution scheme after the take over by Banco Santander.

The EU CoCo market has grown closely to €160 bn outstanding by the mid of 2018. The outstanding CoCos can also be classified in different groups depending on their characteristics. At the start of the CoCo time period, the conversion CoCos were getting more attention due to the recovery involved for the investor if the share would appreciate after the point of crisis. Due to mandates of institutional investors, who are not allowed to hold equity, the write-down CoCos are more marketable. Furthermore, write-down CoCos are regarded as more transparent. We witnessed in 2013 more issuance of write-down CoCos compared to conversion CoCos (De Spiegeleer et al. 2014). Notice that the write-down CoCos are also specifically interesting for financial institutions that are not-listed on a stock exchange.

One should remark that CoCos are not allowed for broad-based bond indices. However the Bank of America Merrill Lynch was first to publish a CoCo index in order to provide a benchmark for the CoCo markets returns. This index tracks the performance of investment and sub-investment grade CoCo bonds since January 2014. The index contains CoCos from major domestic and eurobond markets. In June 2014 also Barclays followed by the launch of a new CoCo index which contained 65 CoCos both conversion and write-down CoCos. Also Credit Suisse and Markit provide different CoCo indices. In May 2018 the first CoCo ETF was launched by WisdomTree.

1.8 Conclusion

Contingent Convertible bonds have their roots in the financial crisis of 2007–2008. These hybrid bonds are constructed to provide extra capital for a distressed bank while keeping it in a going concern. Due to coupon cancellation of the CoCo after a trigger event, they automatically decrease payments in times of stress. In context of Basel III requirements, these loss-absorbing instruments are given an important place. CoCos can get the incentives of the management back on track. Before the crisis there was a reputation of too-big-to-fail with the idea that the regulator would support large failing banks with taxpayers' money. Today, this is no longer that much the case as banks have more loss absorbing capital to survive these crisis on their own. Furthermore, the dilution effect for the existing shareholders after conversion of a CoCo, can set the incentive of keeping the regulatory capital above trigger levels.

Some drawbacks have been discovered and explained while breaking down the construction of a CoCo. Comments are given to the definition of the triggering. Some argue the use of a quarterly available accounting ratio. They state that this ratio is backward looking and might be tripped too late. Also the time point of a regulator intervening with use of CoCo non-viability trigger is highly unclear and can be based on regulatory non-public available information. Other concerns go to the contagion effects if a major CoCo conversion or write-down takes place. One source of the financial crisis has been addressed to the high interconnectedness between financial institutions. This problem is not reduced by issuing CoCos if other financial institutions are allowed to invest. At last, the death spiral effect explains the liquidity problems for the investors hedging their holdings in CoCos leading to a further decrease of share prices.

In the next chapter a detailed description is given of multiple pricing models for CoCos. Applying these models, creates insights on different aspects such as the sensitivity of CoCo bonds to underlying market drivers, their credit rating and the distance of triggering.

Chapter 2
Pricing Models for CoCos

Each CoCo bond is different and this lack of standardisation proves to be a real challenge. Also comparing CoCo bonds of different banks against each other is not straightforward. The actual valuation of a CoCo incorporates the modeling of both the trigger probability and the expected loss for the investor. Some would argue that modelling contingent debt is an impossible task. After all, how could one possibly model an accounting trigger taking place or a regulator pulling the non-viability trigger on a CoCo bond? The only CoCos for which an adequate financial model could deliver an acceptable theoretical price would be those with a market trigger. In such a case, the loss absorption mechanism is activated as soon as an observable variable such as for example a share price level or a credit default swap spreads drops below a specified barrier. However none of such CoCo bonds have been issued so far (De Spiegeleer et al. 2014).

There exist multiple models for the valuation of a CoCo and each one has its own assumptions and drawbacks. The most simplified models which can be derived with a back of the envelop calculation are classified under the term heuristic models. One should not be surprised that the pricing of a sophisticated instrument based on heuristic models will be inaccurate. An example of a rule-of-thumb model to price CoCo bonds is the yield based method. With this approach the capital hierarchy of the liabilities is used in order to derive an estimate of the yield. Once the value of the yield is given, the CoCo price follows from discounting all future coupon payments with the derived yield. As explained in previous chapter, the ranking of CoCos on the balance sheet can be cumbersome especially near a stress event. Furthermore, a lot of financial market information is needed in order to derive the yield of the CoCo. At last, different interpolating schemes might result in different CoCo prices. As follows, we see that the method leads to a straightforward approach although there is a lack of accuracy.

A more evolved approach is based on structural models. A structural model or firm-value model sets the asset price of the issuing company as the stochastic parameter driving the value of bonds or equity. The asset price of a bank is unfortunately an unobservable parameter. Every quarter when the bank reports its earnings, it discloses the value of its assets and provides for a full breakdown of these assets on the balance

J. De Spiegeleer et al., *The Risk Management of Contingent Convertible (CoCo) Bonds*, SpringerBriefs in Finance, https://doi.org/10.1007/978-3-030-01824-5_2

sheet. But in between reporting dates, the asset prices remain unknown. The only observable real-time information consists of share prices, bond prices, credit default swaps and equity option prices. A second difficulty when applying structural models in the CoCo bond valuation is the fact that one needs to model the regulatory capital of the bank in stead of the economic capital. Structural models are often based on the theory of Merton and can be found in e.g. Pennacchi (2010), Fitch Solutions (2011), Pennacchi et al. (2014), Brigo et al. (2015) and Albul et al. (2013). A further extension to the classical Merton model is investigated in Madan and Schoutens (2011). In this paper not only the assets but also the liabilities are assumed to be risky. The authors derive a fundamental model for the CoCo price using conic finance techniques.

With structural models the moment of triggering is based on performance of the assets. However, we can estimate the point of triggering also from other perspectives. The next chapters focus on market implied models. For these models the derivation is based on market data such as share prices, credit default spreads and volatilities. Since CoCos are hybrid instruments, with characteristics of both debt and equity, different approaches do exist for pricing CoCos. First a rather simplistic model is introduced that looks at the CoCo like a credit investment. Afterwards a more sophisticated market implied model is derived, called the equity derivatives approach which will break down the CoCo bond in different exotic options. Both models presented here closely follow the work of De Spiegeleer and Schoutens (2012a) and De Spiegeleer et al. (2014). The equity and credit derivatives approach were introduced in a Black–Scholes framework. Pricing CoCos under smile conform models can be found in Corcuera et al. (2013). Further extensions and discussions can be found in De Spiegeleer and Schoutens (2012b), De Spiegeleer and Schoutens (2013), Cheridito and Zhikai (2013), De Spiegeleer and Schoutens (2014), Corcuera et al. (2014), Liberadzki and Liberadzki (2016) and Chung and Kwok (2016). A new model, called the implied CET1 volatility model, based on the CET1 ratio and its CET1 volatility is added at the end of this chapter.

We note that for simplicity we work here with a simple continuously compound interest rate r. For practical purposes obviously the term structure of interest rates should be taken into account. Modification of the formulas are more or less straightforward but are here presented with a single interest rate to simplify notation. A further step would be to include stochastic interest rates. This is to some extend possible but involves an estimation of parameters driving the stochastic behaviour including potentially some correlation with for example the equity dynamics. This itself is not trivial and has not been incorporated here.

2.1 Credit Derivatives Approach

The credit derivatives approach focuses on the estimation of an extra yield added on top of the risk-free rate as a fair compensation for the risks taken by the investors. This extra yield is called the CoCo spread (cs_{CoCo}) and allows calculating the value of the CoCo. In this approach the CoCo is modeled from a standard debt pricing method using the well-known relationship, called the credit triangle.

2.1.1 Credit Triangle

The credit triangle can be derived from a zero-coupon bond. Assume we have a zero-coupon bond with face value N, maturity T and default intensity λ. The default intensity is the instantaneous probability of default. The time when a default occurs, is unpredictable but can be modeled using a Poisson process with intensity λ. It is well known that the interval times between two Poisson distributed events are independent and exponentially distributed with the same parameter. The probability that no default occurs over an interval of length T can be measured as the probability that the first interval time is larger than T. Based on the exponential distribution, we derive the following equation for the survival probability (p_s):

$$p_s = \exp(-\lambda T) \approx 1 - \lambda T \tag{2.1}$$

Without a default event, the investor receives the initial value (N) at maturity. If during the life of the bond there has been a default, only a recovered part of the initial value will be paid out at maturity. The recovered value is denoted by πN with recovery rate $\pi \in [0, 1]$. The bond's expected value becomes the sum over all discounted payoff values times the probability of receiving this payoff at T. An approximation for the zero-coupon bond is hence given in terms of the default intensity by:

$$B = e^{-rT}[p_s N + (1 - p_s)\pi N] \tag{2.2}$$

$$\approx Ne^{-rT}[1 - \lambda(1 - \pi)T] \tag{2.3}$$

with π the recovery rate from a trigger event. Notice that the value of a zero-coupon bond can also be expressed in terms of its credit spread (c):

$$B = Ne^{-(r+c)T} \approx Ne^{-rT}[1 - cT] \tag{2.4}$$

Combining Eqs. 2.3 and 2.4 leads to the credit triangle:

$$c = \lambda(1 - \pi) \tag{2.5}$$

This well-known rule of thumb gives a relation between the credit spread, default intensity and recovery rate and will be applied in the credit derivatives approach to price a CoCo.

2.1.2 CoCo Pricing

Analogue to a default event in corporate debt, the CoCo has the trigger event. The credit triangle and pricing method of a bond can easily be applied in the pricing of a CoCo. Similar to the probability of default in a zero-coupon bond, the probability of a trigger event (p_T) of a CoCo bond before T years can be expressed in terms of a trigger intensity (λ_T) as follows:

$$p_T = 1 - \exp(-\lambda_T T) \tag{2.6}$$

Based on previous equation one can now calculate the trigger intensity as:

$$\lambda_T = -\frac{1}{T}\log(1 - p_T) \tag{2.7}$$

Notice that this trigger intensity should naturally be higher than the default intensity of a corporate debt because the probability of a trigger event is larger than the probability of a default.

The formula for the credit derivatives approach is now derived from the credit triangle (Eq. 2.5). From the credit spread of a CoCo with recovery rate (π), an implied trigger level can be derived and vice versa by:

$$cs_{CoCo,T} = -\frac{(1-\pi)}{T}\log(1 - p_T) \tag{2.8}$$

The CoCo spread ($cs_{CoCo,T}$) together with the risk free interest rate results in the yield (y_T):

$$y_T = cs_{CoCo,T} + r_T \tag{2.9}$$

where r_T denotes the risk free rate for a term of T years. In the last step, the price of the CoCo can be obtained, by discounting the cash-flows with the corresponding yield:

$$P = \sum_i c_i e^{-y_{t_i} t_i} + N e^{-y_T T} \tag{2.10}$$

where c_i are the future coupon payments paid at times t_i and N the face value.

2.1.3 Recovery Rate

The recovery rate (π) of a conversion CoCo can be derived from the conversion price. At the moment of conversion, the investor of a conversion CoCo receives C_r shares with a market value denoted with S^*. The loss of a conversion CoCo in case of a trigger is defined as the initial value minus the total value of the shares the investor has received from the conversion into equity (see also Fig. 2.1):

$$\begin{aligned} L &= N - C_r S^* \\ &= N\left(1 - \frac{S^*}{C_p}\right) \end{aligned}$$

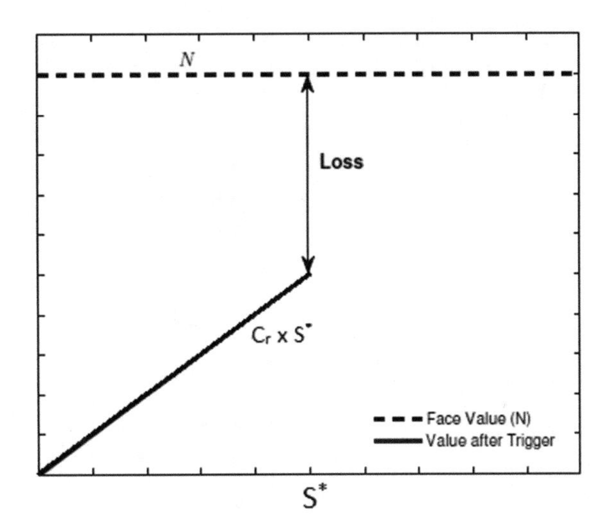

Fig. 2.1 Loss of a CoCo with conversion to equity

Hence the recovery of the conversion CoCo is defined by $\pi = S^*/C_p$. For a write-down CoCo, the loss is equal to the fraction of the notional that will be written down as defined in the prospectus.

2.1.4 *Probability of Triggering*

The only non-trivial value in the pricing formula (Eq. 2.10) is an estimate of the probability of triggering. This is by far the most challenging task in pricing a CoCo. One way to go around the problem is to associate the triggering of the CoCo with the event of the share price dropping below a barrier level S^* as considered in De Spiegeleer and Schoutens (2012a). This level corresponds to the share price at the moment the CoCo bond gets triggered and will be called the (implied) trigger level. Hence a trigger event on the balance sheet is replaced by an event for the share price level as illustrated in Fig. 2.2 for an accounting trigger.

Different stock price models like Black–Scholes, Heston, CEV, Variance-Gamma etc. can be used to find the probability of breaching the trigger level. For example the probability that the share price drops below S^* under the Black–Scholes model is given by the first exit time equation:

$$p_T^* = \Phi\left(\frac{\log(S^*/S) - \mu T}{\sigma\sqrt{T}}\right) + \left(\frac{S^*}{S}\right)^{2\mu/\sigma^2} \Phi\left(\frac{\log(S^*/S) + \mu T}{\sigma\sqrt{T}}\right) \tag{2.11}$$

where

$$\mu = r - q - \frac{\sigma^2}{2}$$

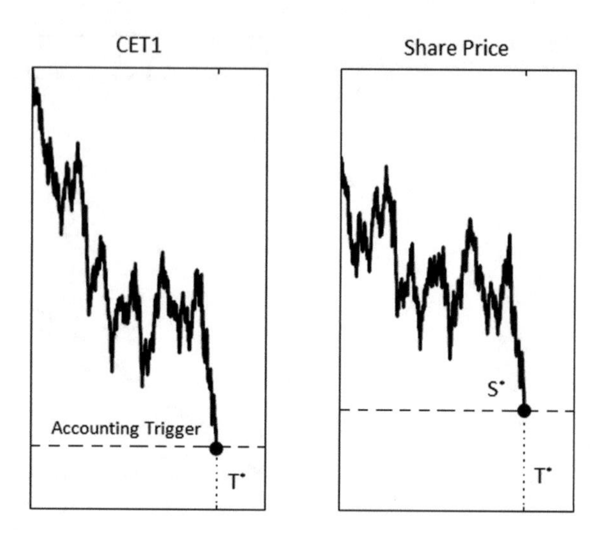

Fig. 2.2 An accounting trigger is modeled as the stock price dropping below trigger level S^*

with the risk-free interest rate r, dividend yield q, the current share price S, the volatility parameter σ and

$$\Phi(x) = \text{Probability}(X \leq x) \tag{2.12}$$

is the probability that a standard normal random variable X takes a value smaller than x.

This method is easy to implement but it uses some simplifications. The largest shortcoming of the credit derivatives method is that the financial losses coming from the cancelling in the coupon stream are not taken fully into account. Therefore the credit derivatives approach is rather a rule of thumb. It is an easy pricing method but it uses some simplifications. We move on to a more accurate approach in the next section.

2.2 Equity Derivatives Approach

Notice that the trigger is defined by a particular CET1 level or decided upon a regulator's decision. Since these trigger mechanisms are hard to model or even to quantify, we project the trigger into the stock price framework. The event of an accounting ratio being hit, is assumed to be equivalent to the share price dropping for the first time below S^*. Hence we assume the existence of a barrier share price level S^* such that the moment when the CET1 ratio fails to stay above the minimum trigger level, coincides with the share price level dropping below S^*. This level corresponds actually to the value of the underlying share on the moment the bond is triggered into conversion or is written down.

As a result the valuation of a CoCo bond is transformed into a barrier-pricing exercise in an equity setting. Under the stock price framework the CoCo bond can be broken down to several different derivative instruments. In first place the CoCo behaves like a standard (non-defaultable) corporate bond where the holder will receive coupons c_i on regular time points t_i together with the principal N at maturity T. However, this bond should be knocked out if the stock level drops below the barrier level S^*.

This first component is the zero-coupon corporate bond (**ZC**). The second component (**Cpn**) is a sum of the Binary-Down-and-Out (BDO) options. These BDO options payout zero in case the stock drops below a certain value. The BDO options in the CoCo construction have maturities t_i for each coupon c_i. Under Black–Scholes stock price model the explicit formula for these options can be found in Rubinstein and Reiner (1991). The theoretical price under the Black–Scholes model of the full write-down CoCo expiring T years from now, becomes:

$$\begin{aligned}
\mathbf{P} &= \mathbf{ZC} + \mathbf{Cpn} + \Pi \\
&\text{where} \\
\mathbf{ZC} &= \underbrace{N\exp(-rT)}_{\text{zero coupon}} \times \underbrace{\left[\Phi(x_1 - \sigma\sqrt{T}) - \left(\frac{S^*}{S}\right)^{(2\lambda-2)} \Phi(y_1 - \sigma\sqrt{T})\right]}_{\text{Prob. of no trigger before } T} \\
\mathbf{Cpn} &= \sum_{i=1}^{k} \underbrace{c_i \exp(-rt_i)}_{\text{present value of } c_i} \times \underbrace{\left[\Phi\left(x_i - \sigma\sqrt{t_i}\right) - \left(\frac{S^*}{S}\right)^{(2\lambda-2)} \Phi\left(y_i - \sigma\sqrt{t_i}\right)\right]}_{\text{Prob. that at time } t_i \text{ the trigger level } S^* \text{ was never reached}} \\
\Pi &= \text{Recovery value}
\end{aligned} \tag{2.13}$$

$$\begin{aligned}
&\text{with} \\
x_1 &= \frac{\log(S^*/S)}{\sigma\sqrt{T}} + \lambda\sigma\sqrt{T} \\
y_1 &= \frac{\log(S/S^*)}{\sigma\sqrt{T}} + \lambda\sigma\sqrt{T} \\
x_i &= \frac{\log(S^*/S)}{\sigma\sqrt{t_i}} + \lambda\sigma\sqrt{t_i} \\
y_i &= \frac{\log(S/S^*)}{\sigma\sqrt{t_i}} + \lambda\sigma\sqrt{t_i} \\
\lambda &= \frac{r-q+\frac{1}{2}\sigma^2}{\sigma^2} \\
k &= \text{Total number of coupons} \\
\sigma &= \text{Volatility parameter} \\
t_i &= \text{Time until the } i \text{ th coupon payment} \\
T &= \text{Time until maturity}
\end{aligned} \tag{2.14}$$

Remark that this pricing formula includes a parameter λ as defined in Eq. 2.14. The parameter might not be confused with the default intensities in previous section.

The recovery value, denoted with Π, equals zero for a full write-down CoCo. For a fractional write-down CoCo the recovery rate is given in the CoCo's prospectus. In case of a conversion CoCo, we remark that the investor receives an amount of shares at the triggering. The number of shares is denoted by the conversion ratio C_r. Hence, this part can be modeled as C_r down-and-in asset-(at hit)-or-nothing options. For a conversion CoCo, we hence have also a non-zero recovery (Π) in Eq. 2.13:

$$\begin{aligned}
\Pi &= C_r \times S^* \left[\left(\tfrac{S^*}{S}\right)^{a+b} \Phi(z) + \left(\tfrac{S^*}{S}\right)^{a-b} \Phi(z - 2b\sigma\sqrt{T}) \right] \\
\text{with} & \\
z &= \tfrac{\log(S^*/S)}{\sigma\sqrt{T}} + b\sigma\sqrt{T} \\
a &= \lambda - 1 \\
b &= \sqrt{a^2 + \tfrac{2r}{\sigma^2}}
\end{aligned} \tag{2.15}$$

Some argue that even a conversion CoCo will have a recovery value equal to zero. This can be explained by the fact that the CoCo will trigger in a life-threatening situation and hence the number of shares received will be worthless. This was actually confirmed by the Banco Popular trigger in June 2017. However high trigger CoCos are assumed to absorb losses in a going-concern. For these CoCos the issuer will probably recover from its stress event.

The (unknown) trigger level S^* can be implied from the market price of the Contingent Convertible bond. Further, also the value of a new CoCo can be found based on the market price of a similar outstanding CoCo. One could derive the implied barrier of the outstanding CoCo from its market price and use the same share price barrier level in the pricing of the new CoCo. Notice also that if a bank has several CoCo bonds outstanding all sharing the same accounting trigger, these bonds should have the same implied trigger level S^*. This way over- or under valuated CoCos can be detected. For example, assume the implied trigger level of one CoCo is higher compared to other CoCos with similar accounting trigger and the same issuer. This means the CoCo market price assumes this CoCo will be triggered before the other CoCos which is not possible. Hence this CoCo is under valued.

Besides the discussion point on the recovery for a conversion CoCo, another hindrance is the lack of knowledge regarding the volatility parameter σ. An approach to find an estimate for σ was proposed by JP Morgan (Morgan 1999). For equity derivative markets, usually the implied volatility is available for options with rather short time to maturity and with strikes around the current spot price. However, a CoCo bond has typically a long maturity usually 5 years or more and their conversion takes place in a stress event, when stock trades at low levels. In the calibration of pricing models we will include derivatives with a similar characteristics of high maturity and low strikes. An heuristic way to find an estimate for σ can be found based on a particular credit default swap (CDS) level. We see a CDS contract with zero recovery as a deep out-the-money (OTM) put option. This way we can find the implied volatility parameter for which the OTM put option matches with the market spreads of a zero recovery CDS. Hence in this approach we actually use the CDS spreads as an input to find the unknown parameter in the pricing model. In Chap. 4, the impact of skew on the pricing of CoCos will be investigated based on the pricing model of Heston.

2.3 Implied CET1 Volatility Model

The most straightforward model without any stock price model involved is the following implied CET1 volatility model. Despite the fact that one knows the value of the CET1 ratio only on a quarterly basis, we model this ratio as a continuous geometric Brownian motion in the absence of any drift. This model is referred to as Black's model:

$$\frac{d\text{CET1}_t}{\text{CET1}_t} = \sigma_{\text{CET1}} dW_t \tag{2.16}$$

The probability p_T^* that this accounting ratio hits a trigger level during a time horizon T is given by the following equation (Su and Rieger 2009):

$$p_T^* = \Phi\left(\frac{\log\left(\frac{\text{Trigger}}{\text{CET1}}\right) - \mu T}{\sigma_{\text{CET1}}\sqrt{T}}\right) + \left(\frac{\text{Trigger}}{\text{CET1}}\right)^{-1} \Phi\left(\frac{\log\left(\frac{\text{Trigger}}{\text{CET1}}\right) + \mu T}{\sigma_{\text{CET1}}\sqrt{T}}\right) \tag{2.17}$$

with

μ	$=$	$-\frac{\sigma^2_{\text{CET1}}}{2}$
Trigger	:	Contractual CET1 trigger level
σ_{CET1}	:	Volatility of the CET1 ratio
T	:	Maturity of the contingent convertible
CET1	:	Current CET1 ratio (CET1_0)

The equation above is similar to the first exit time equation (Eq. 2.11). In previous sections we modeled the stock price dropping below the share price trigger level. Here it gives the probability that a CET1 ratio will touch the accounting trigger somewhere between today and the expiration of the bond T years from now.

Introducing the trigger distance $D = \frac{CET1}{\text{Trigger}}$ into Eq. 2.17 results in:

$$p_T^* = 1 - \Phi\left(\frac{\log(D) + \mu T}{\sigma_{\text{CET1}}\sqrt{T}}\right) + D\Phi\left(-\frac{\log(D) - \mu T}{\sigma_{\text{CET1}}\sqrt{T}}\right) \tag{2.18}$$

This equation expresses the probability that the trigger is going to take place. From p_T^* we can now determine the value of the CoCo spread ($cs_{CoCo,T}$):

$$cs_{CoCo,T} = -\frac{\log(1 - p_T^*)}{T} \times (1 - \Pi_{\text{CoCo}}) \tag{2.19}$$

with Π_{CoCo} the recovery ratio after a trigger event.

The equation above is an extension of the credit derivatives method. Using Eq. 2.19 to price a particular contingent convertible is not straightforward. The main hindrance is the lack of knowledge regarding the recovery ratio and the level of the volatility of the CET1 ratio. In the case of a write-down CoCo bond where $\Pi_{\text{CoCo}} = 0$, this will still leave us with the need to determine the value of σ_{CET1} in order to determine the value of the contingent convertible. However, the other way around is more interesting. From the CoCo market prices, the model can be applied to find an implied volatility level for the CET1 ratio. This approach is very similar to derivation of an implied volatility from the vanilla option prices in the equity derivatives markets. Here we will look at the CoCo bond as kind of a derivative instrument with the CET1 level as underlying. From the given CoCo market prices we infer the corresponding implied CET1 volatility such that our model price matches with the market price.

A more sophisticated model in the same spirit can be derived from the equity derivatives approach where the exotic options are knocked out if the CET1 level drops below the trigger ($D < 1$). The theoretical price of the full write-down CoCo and expiring T years from now, becomes:

$$
\begin{aligned}
\mathbf{P} &= \mathbf{ZC} + \mathbf{Cpn} \\
\mathbf{ZC} &= \underbrace{N \exp(-rT)}_{\text{zero coupon}} \times \underbrace{(1 - p_T^*)}_{\text{Probability of no trigger before } T} \\
\mathbf{Cpn} &= \sum_{i=1}^{k} \underbrace{c_i \exp(-rt_i)}_{\text{present value of } c_i} \times \underbrace{\left[\Phi\left(x_i - \sigma_{\text{CET1}}\sqrt{t_i}\right) - D\Phi\left(y_i - \sigma_{\text{CET1}}\sqrt{t_i}\right)\right]}_{\text{Prob. that at time } t_i \text{ the CET1 trigger was never reached}} \\
&\text{with} \\
x_i &= \frac{\log(D)}{\sigma_{\text{CET1}}\sqrt{t_i}} + \frac{\sigma_{\text{CET1}}\sqrt{t_i}}{2} \\
y_i &= -\frac{\log(D)}{\sigma_{\text{CET1}}\sqrt{t_i}} + \frac{\sigma_{\text{CET1}}\sqrt{t_i}}{2} \\
k &= \text{Total number of coupons}
\end{aligned}
\tag{2.20}
$$

As mentioned above, the merit of the equations above resides with the fact that starting from the value of a CoCo bond, or its spread cs_{CoCo}, one can determine the $\overline{\sigma}_{\text{CET1}}$. This is the implied volatility of the CET1 ratio. It reflects the view of the market on the dynamics of the CET1 ratio under the assumption of Eq. 2.16. We will illustrate the applicability of this model on one of the contingent convertibles issued by Credit Suisse.

Example:
On October 21, 2015, Credit Suisse reported its capital ratios for the third quarter of 2015, as of September 30, 2015. The fully phased-in CET1 ratio was reported by the bank to be equal 10.2%. Given the fact that the Tier 2 CoCo bond has a 5% trigger, we have that $D = 2.04$.

In order to calculate $\overline{\sigma}_{\mathrm{CET1}}$ for Credit Suisse, we use the CoCo price on the publication day of the CET1 ratio. Because we are dealing with a full write-down CoCo, Π_{CoCo} is set equal to zero in Eq. 2.19. Using this equation, the corresponding implied volatility for the CET1 ratio is 18.88%.

CREDIT SUISSE	
Name	CREDIT SUISSE 5.75%
ISIN	XS0972523947
Issue Date	18-Sep-13
CET1 Report Date	21-Oct-15
CET1 Fully phased-in	10.2
Trigger	5
CoCo Price	112.88
CoCo Spread	275

2.4 Conclusion

In the credit derivatives approach the CoCo is modeled from a standard debt pricing method where the CoCo spread, i.e. the extra yield on top of the risk free yield, is derived from the well-known credit triangle relationship. One should remark that no coupon cancellation is really taken into account. A more complex approach is the equity derivatives approach. The time of trigger is modeled as the stock price dropping below a certain barrier level. Within this approach the CoCo bond is split in a standard bond and different exotic barrier options. A last described CoCo model applies Black's model to the CET1 ratio. Based on this model, new information can be found from the CoCo market price about the implied CET1 volatility. Each of the three models described here will be used in the next chapters to capture and manage the risks within CoCo bonds.

Chapter 3
Sensitivity Analysis of CoCos

Knowing and understanding the exposures of different financial instruments is a crucial part of the investment process. These insights are necessary to provide the investor an overview which factors have the largest influence on the price. Once the holder of the financial instrument knows these sensitivities, he/she can start investing in other financial instruments to decrease the overall risk exposure, if he/she finds this appropriate. This procedure is referred to as hedging. In this chapter a sensitivity analysis is provided to explain how moves in the theoretical CoCo price are driven by changes in other market values including the underlying stock, the credit spread and the interest rates.

The Greek parameters explain the sensitivity of the derivatives price. The concept of the Greeks is defined for option prices but can be extended to other derivatives prices such as CoCo bonds. Explicit formulas for the Greek parameters (Theta, Gamma, Vega and Rho) of exotic barrier options can be found. However these derivations are error prone and time consuming. They require also specific assumptions regarding the distribution of the price. Therefore the Greeks of a CoCo are here approximated using the finite difference method.

An investor can manage or hedge these Greeks by reducing the exposure to the underlying stock price and credit risk. The hedging strategy requires multiple investments per different CoCo issuer in the portfolio. Instead of measuring the sensitivity of each individual CoCo with respect to their own specific underlying, we measure their sensitivity with respect to a general equity index (Eurostoxx-50 or FTSE-100) and a credit index (iTraxx). Based on these sensitivities one might hedge a full CoCo portfolio of multiple different issuers, with investments in only a few market indices. This results in a more reasonable and practical approach to hedge CoCos. The exposures for CoCos from different issuers, are hence combined in order to reduce the hedging costs.

J. De Spiegeleer et al., *The Risk Management of Contingent Convertible (CoCo) Bonds*, SpringerBriefs in Finance, https://doi.org/10.1007/978-3-030-01824-5_3

3.1 Hedging CoCos

CoCo bonds are developed to absorb losses in times of crisis. When the stock price drops deeply, it is more likely to lead to a trigger event for the CoCo bond. The CoCo can get converted or written down which results in large losses for the CoCo bond holder. Each investor of CoCos should be aware of this risk and might want to hedge his position. In times of stress, one may just want to close the positions in order to avoid further losses. However in such distressed periods the liquidity of CoCos can be significantly reduced and hence selling positions could be only possible at fire-sale-prices. Therefore, one is maybe forced to look for alternative ways to reduce risk in the positions one can not exit. Hedging, by taking positions in other instruments, can in these circumstances be a way to proceed.

The classical approach to hedge the risk for falling stock prices is an investment in put options. A put option is designed to create value for falling markets or provide protection against them. A straightforward solution for protection of the CoCo holder is investing in deep out-the-money (OTM) put options. This hedging strategies can be based on market-making values. One might invest in the put options with the lowest strike available on the market, say around 50–60% of the current stock price. The number of put options bought can for example be based on investment restrictions. For example, assume the coupons are set at 5% and the notional equals one million EUR. The coupon payments will be 50,000 EUR. A limit is set on the amount invested in the puts. The investment in puts might as such be limited to 20% of the coupon value, equal to 10,000 EUR. Another solution is to derive the number of put options from a limit on the losses. One could for example buy the number of put options such that the total loss of this strategy is say below 30% of the notional of the CoCo. Remark that the available put options in the market have typically a short term maturity compared with the CoCo bonds maturity. To cover the full long term of the CoCo, the positions need to be rolled over when reaching their maturity. For example, the put options have a maturity around one year and are rolled over in case the term becomes less than 3 or 4 months.

The hedging strategy with put options has three different limitations. First, these hedging strategies require a separate investment for each different CoCo issuer in the CoCo portfolio. Since the CoCos can have different issuing institutions, different underlying stocks can arise in the portfolio. This means that an investment in multiple put options is necessary. Second, there is no perfect hedge between the CoCo and the put option. The limit on the losses clearly depends on the stock price in a stress event. It is possible to buy the number of put options in order to have the return of the puts match the notional in case the stock price drops to zero. However there will be losses in case the CoCo is triggered and the put option matures out the money. Last, the cost of the hedge can change on a daily basis. In the hedging explained in next sections, a solution is provided based on the sensitivity parameters of the CoCo bond.

3.2 Sensitivity Parameters

A more accurate but model-driven approach is a hedging strategy based on the sensitivity parameters for CoCos. These sensitivity parameters are called the Greeks since each parameter is represented by a letter of the Greek language alphabet. The parameters are the mathematical derivatives of the CoCo price with respect to different underlying drivers. For some models (e.g. our EDA model), these derivatives can be found explicitly from a closed form formula of the CoCo price. This approach has the disadvantage that it depends heavily on the underlying pricing model. The flaws in our approach will be displayed by comparing the CoCo market performances with the estimated performance in the CoCo from the sensitivity parameters and moves in the underlying market drivers.

3.2.1 The Greeks

The Greeks refer to sensitivity parameters describing the sensitivity of an option price with respect to different underlying market drivers. In mathematical terms, the Greeks are the partial derivatives with respect to the underlying stock price, interest rate, etc. In practice, traders look at the Greeks as exposures of the option and understand how they will dynamically change over time or when the market moves.

The CoCo is a collection of exotic options under the assumptions of the EDA pricing model of Chap. 2. The exposures of a CoCo can hence be described in terms of its Greeks. In this chapter we describe four Greeks. The first and second order sensitivity of the CoCo price with respect to the underlying share price (S) are called delta ($\boldsymbol{\Delta}$) and gamma ($\boldsymbol{\Gamma}$) respectively. The sensitivity with respect to the interest rate curve (r) is expressed by the Greek rho ($\boldsymbol{\rho}$). Last we define the Greek vega ($\boldsymbol{\nu}$). This Greek originally measures the sensitivity with respect to the volatility parameter (σ) used in the Black–Scholes model. In the implementation of the EDA, we derived the volatility parameter from the CDS spreads. As such we define here the vega as a measure of the sensitivity with respect to the credit spread (c). Hence the Greeks defined in this chapter measure sensitivity with regard to the input variables of the EDA pricing model.

In this chapter the Greeks are defined by:

$$
\begin{aligned}
\Delta &= \frac{\partial P}{\partial S} \\
\Gamma &= \frac{\partial \Delta}{\partial S} = \frac{\partial^2 P}{\partial S^2} \\
\boldsymbol{\nu} &= \frac{\partial P}{\partial c} \\
\boldsymbol{\rho} &= \frac{\partial P}{\partial r}
\end{aligned}
\qquad \text{with} \qquad
\begin{aligned}
P &= \text{Model CoCo price} \\
S &= \text{Stock price} \\
c &= \text{Credit spread} \\
r &= \text{Interest rate}
\end{aligned}
$$

The Greeks are useful tools to derive in which instruments we should invest to diminish the total risk in our portfolio. Delta denotes how many shares minimise the sensitivity of the total portfolio with share price moves. The gamma denotes the hedging error introduced by the delta ratio. Hence the value of gamma will tell us how often we need to rebalance or how much error can be included when not repositioning on time. The Greek rho is more difficult to translate in a hedge since one cannot trade the interest rate directly. In order to hedge these types of risk, the investor should trade through interest rate instruments with an opposite ρ. To reduce vega, one could invest in credit spread derivatives (Leoni 2014).

3.2.2 Estimating the Greeks of a CoCo

In this section, the risks in CoCo bonds are hedged as if a CoCo bond is a portfolio of a standard bond structure and multiple barrier options as described by the EDA pricing model. The Greeks of the CoCo can now be derived from the mathematical derivatives of the prices of the barrier options. However in practice, calculating the derivatives manually is a time consuming and error prone work. A more easy and practical way is by approximating the derivatives by first or second order finite differences. In case the CoCo price models have been implemented, this is just a task of evaluating the model for different input values like stock price, interest rates, volatilities etc.

Denote the model CoCo price by $P(S, r, c)$ when the current underlying share price is S, the underlying interest rate equals r and the credit spread is c. In this chapter, each exact premium price of a CoCo is transformed to a quote value. We divide the premium with its notional N and multiply this amount with 100 to obtain the quote. As such all CoCo quotes are situated around 100. Furthermore, the quote value has no unit or currency whereas the exact premiums are expressed in a certain CoCo currency.

The Greeks are approximated based on the finite difference approximation as follows:

$$\boldsymbol{\Delta} \approx \frac{P(1.01S, r, c) - P(S, r, c)}{0.01S} \tag{3.1}$$

$$\boldsymbol{\Gamma} \approx \frac{P(1.01S, r, c) - 2 * P(S, r, c) + P(0.99S, r, c)}{(0.01S)^2} \tag{3.2}$$

$$\boldsymbol{\nu} \approx \frac{P(S, r, c + 0.0001) - P(S, r, c)}{0.0001} \tag{3.3}$$

$$\boldsymbol{\rho} \approx \frac{P(S, r + 0.0001, c) - P(S, r, c)}{0.0001} \tag{3.4}$$

For delta and gamma, the price model is evaluated for the exact market stock price and for a one percent higher or lower stock price. To derive the Greek rho, the model price is derived for the current market zero rate curve and when this interest rate

Table 3.1 The theoretical CoCo Quote (in bold) on September 1, 2015 for different underlying stock prices, an increased credit spread by one basispoint or a shift in the interest rates by one basispoint. The corresponding values of the underlying equity market are displayed and are increased or decreased by one percentage

CoCo Name	S	$1.01 * S$	$0.99 * S$	CS	IR
SOCGEN (USD)	47.26	47.73	46.78	+1bps	+1bps
SOCGEN 8 1/4 09/29/49	108.47	108.80	108.12	108.34	108.44
SOCGEN 7 7/8 12/29/49	101.97	102.51	101.42	101.72	101.93
LLOYDS (GBP)	0.7556	0.7632	0.7480	+1bps	+1bps
LLOYDS 7 7/8 12/29/49	106.54	107.09	105.99	105.95	106.48
LLOYDS (USD)	1.158	1.170	1.146	+1bps	+1bps
LLOYDS 7 1/2 04/30/49	105.46	105.94	104.97	105.06	105.41
CS (USD)	25.40	25.65	25.14	+1bps	+1bps
CS 7 7/8 02/24/41	104.27	104.37	104.17	104.23	104.26
CS (CHF)	24.43	24.67	24.18	+1bps	+1bps
CS 7 1/8 03/22/22	105.45	105.63	105.26	105.38	105.44

curve is horizontal shifted upwards by one basispoint. At last, the impact of different credit spreads is measured by increasing the credit spread curve by one basispoint.

These definitions of delta and gamma express the first and second order change in CoCo quotes for an increase in the stock price by one currency unit. For example if the underlying equity increases by 1 USD, the CoCo price quote will increase by Δ. The Greeks explaining the impact in terms of percentage change or returns of the stock instead of unit change, are:

$$\mathbf{\Delta_r} \approx \frac{P(1.01S, r, c) - P(S, r, c)}{0.01} \tag{3.5}$$

$$\mathbf{\Gamma_r} \approx \frac{P(1.01S, r, c) - 2 * P(S, r, c) + P(0.99S, r, c)}{(0.01)^2} \tag{3.6}$$

Assume the underlying equity increase by 1%, then the CoCo price will increase by $\mathbf{\Delta_r}$ times 1%.

As an example we investigate the impact of equity, interest rates and credit spread (volatility) on 6 different CoCos from 3 different issuers. We explain the results here for the SOCGEN 8 1/4 09/29/49 CoCo issued on September 6, 2015. This instrument has a USD currency and pays out a yearly coupon of 8.25%. The dirty CoCo price quote equals 108.47 at September 1, 2015. The underlying stock price equals 47.26 USD. For an increase of the stock price by one percent to 47.73 USD, the CoCo price under the EDA model increases to 108.80. Similarly shifts are given to the credit spread and interest rate by one basispoint. The impact of small changes in the input values of the pricing model on the CoCo quote are displayed in Table 3.1.

The moves in the CoCo quote do coincide with the construction of this instrument. The CoCo price does increase when the underlying stock price increases since this implies moving away from the trigger barrier and hence a decreased conversion risk.

Table 3.2 Greeks of CoCos on September 1, 2015

CoCo Name	Δ	Δ_r	Γ	Γ_r	ν	ρ
SOCGEN 8 1/4 09/29/49	0.7125	33.67	–0.0393	–87.75	–1306	–247.6
SOCGEN 7 7/8 12/29/49	1.145	54.12	–0.0425	–94.80	–2519	–411.9
LLOYDS 7 7/8 12/29/49	72.65	54.89	–141	–80.32	–5874	–601.2
LLOYDS 7 1/2 04/30/49	41.17	47.68	–61.7	–82.75	–3992	–470.5
CS 7 7/8 02/24/41	0.3921	9.957	–0.0595	–38.39	–362.7	–93.93
CS 7 1/8 03/22/22	0.7373	18.01	–0.0970	–57.85	–720.0	–142.1

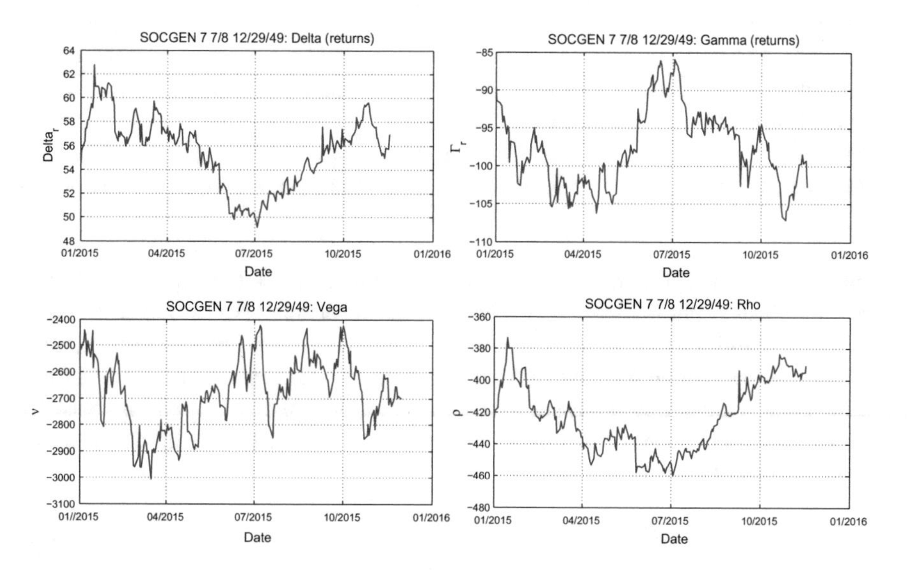

Fig. 3.1 Historical Greeks of SOCGEN 7 7/8 12/29/49 (1-Jan-2015–1-Dec-2015)

The CoCo quote is positively related to the underlying stock price. The interest rate is negatively related with the CoCo price. An increase in interest rates, decreases the discount factors. Since the CoCo price is the sum of discounted future cashflows, the CoCo price decreases when the interest rate increases. Increasing the credit spread curve by a percentage leads to a decreased model SOCGEN CoCo quote of 108.34. This is reasonable since higher volatility implies higher probability of hitting the implied barrier and hence a lower CoCo price. Based on Formulas (3.1)–(3.6), we derive the Greeks in Table 3.2. The Greeks of SOCGEN 7 7/8 12/29/49 are shown from January 1, 2015 until December 1, 2015 in Fig. 3.1.

The changes in the CoCo price can now be related with changes in the underlying market drivers. This relation can be described by a Taylor expansion series. A Taylor Expansion for a function $f(x)$ about $x = a$ is given by:

$$f(x) = \sum_{n=0}^{\infty} \frac{f^{(n)}(a)}{n!}(x - a)^n \tag{3.7}$$

with $f^{(n)}(a)$ the nth derivative of the function $f(x)$ evaluated in $x = a$. We apply here the Taylor Expansion to the CoCo price function $P(S, r, c)$. As such the difference in theoretical CoCo price (ΔP) can be estimated by:

$$\Delta P \approx \mathbf{\Delta} \times \Delta S + \frac{1}{2}\mathbf{\Gamma} \times (\Delta S)^2 + \boldsymbol{\nu} \times \Delta \text{CS} + \boldsymbol{\rho} \times \Delta \text{IR} \tag{3.8}$$

with the change in the underlying stock price ΔS, ΔCS denotes the change in the underlying credit spread and the change in interest rates is denoted by ΔIR. In the following examples the daily change in the 5 year risk free interest rate is used as a measure for ΔIR.

Notice that the change in the stock price in Eq. 3.8 is measured on a daily basis and expressed in the currency of the CoCo. The equation can also be expressed in terms of the returns of the underlying stock price ($R_S = \frac{\Delta S}{S}$) by:

$$\Delta P \approx \mathbf{\Delta_r} \times R_S + \frac{1}{2}\mathbf{\Gamma_r} \times R_S^2 + \boldsymbol{\nu} \times \Delta \text{CS} + \boldsymbol{\rho} \times \Delta \text{IR} \tag{3.9}$$

3.3 Beta Coefficient

An investor can manage or hedge the Greeks by reducing the exposure to the underlying stock price and credit risk. This hedging strategy is based on investments in the underlying equity market. However this strategy with the individual Greeks per CoCo requires an investment in the underlying stock market for each different CoCo issuer in the portfolio. With the beta-coefficient the sensitivity can be estimated with respect to some overall stock market index. This will reduce the costs and decrease the follow-up efforts of the hedging strategy for a CoCo portfolio.

The beta coefficient was first applied in the famous Capital Asset Pricing Model (CAPM) of Sharpe (Sharpe 1964) and Lintner (Lintner 1965). This model values an asset by linking the asset's returns and its risk. The risk in an asset can be divided in two categories: diversifiable and non-diversifiable risk. Beta explains the financial asset's systematic risk or non-diversifiable risk. It denotes the asset risk compared to the overall market risk or the volatility of an assets returns compared to volatility of overall market returns. In the CAPM model, the expected return rate on a security is derived from its beta by the following linear equation:

$$E(R) = R_F + \beta * (R_M - R_F) \tag{3.10}$$

where R_F denotes the rate of a risk-free investment and R_M the return of the market (or a market portfolio of assets). Beta is mostly positive since a negative value would denote a counter-cyclical issuer. In that case, the model suggest that riskier investments in general yield higher returns than the risk-free investment.

The beta values denote the degree to which the equity price fluctuates compared to the general market. More precise, it is the percentage change in the security given a one percentage change in a representative market index. These beta values are derived from a regression analysis where the security is the dependent variable and the market index represents the independent variable. The formula for beta becomes:

$$\beta = \frac{Cov(R, R_M)}{Var(R_M)} \tag{3.11}$$

with R the return of the security and R_M the return of the market index. Remark that different periods can be chosen for the returns i.e. daily, weekly,... returns.

The beta for the equity underlying a CoCo bond is calculated with Formula (3.11). We use as a market index the Eurostoxx 50 index for European CoCo issuers and the FTSE 100 index for UK CoCo issuers. The same procedure is also applied to the credit spread. The beta for the credit spread (denoted by β_{CS}) can be derived from the underlying credit spread changes and the change in a CDS spread index.

Example We show the derivation of the beta values for the CoCos in Table 3.1. The outcome is based on historical data from January 1, 2015 until December 1, 2015. We selected the Eurostoxx 50 equity index for the CoCos from Societe General. For the Lloyds CoCo, we compare the daily returns of Lloyds equity with the daily return in the FTSE 100 index. The relation between the specific underlying equity and the index is shown in Fig. 3.2. The beta's for the equity underlying the CoCo (β) are given in the third column in Table 3.3. The beta for the CDS spreads (β_{CS}) is derived from the daily differences in the spread. The daily change in Markit iTraxx Europe Senior 5 year is applied as a corresponding CDS spread index (see Table 3.3). Notice that these CDS beta values are almost the same for CoCos from the same issuer. The difference appears from the currency of the CoCo.

3.4 Goodness-of-Fit

The approximation describing the relation between the change in the CoCo price and the underlying drivers (Formula 3.9) can be rewritten in terms of more general indices. The following approximation holds:

$$\begin{aligned} \Delta P \approx\ & \mathbf{\Delta_r} \times \beta \times R_{ES50} + \mathbf{\Gamma_r} \times (\beta \times R_{ES50})^2 + \boldsymbol{\nu} \times \beta_{CS} \times \Delta \text{iTraxx} \\ & + \boldsymbol{\rho} \times \Delta \text{IR} \end{aligned} \tag{3.12}$$

This equation is however very similar to a regression model for the impact on the CoCo price. As such we can compare the outcome of the Greeks and the beta coefficients with the results of a regression model. The regression model is defined as:

$$\Delta P \approx \alpha_1 \times R_{ES50} + \alpha_2 \times (R_{ES50})^2 + \alpha_3 \times \Delta \text{iTraxx} + \alpha_4 \times \Delta \text{IR} \tag{3.13}$$

with parameters $\alpha_1, \alpha_2, \alpha_3, \alpha_4$. The results of the regression model in Table 3.4, highlight the significant impact of the change in iTraxx on the change in the CoCo

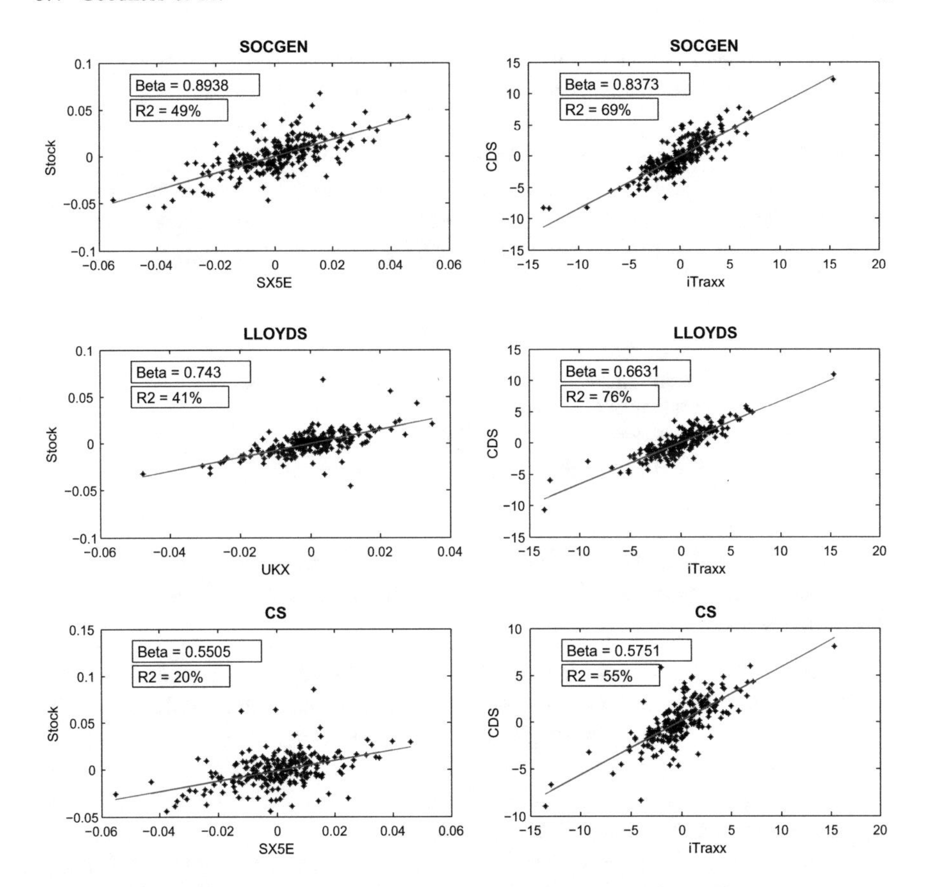

Fig. 3.2 Beta of the equity versus the S&P 500 index (left) and Beta of the CDS spread versus the iTraxx index (right) based on the historical data from 2015

Table 3.3 Beta of daily returns in equity and the beta for the changes in credit spread based on data from January 1, 2015 until December 1, 2015. Note that for Lloyds and CS different betas are obtained for different currencies due to additional FX risk

CoCo	Ccy	Index	β	Rsq (%)	β_{CS}	Rsq (CDS) (%)
SOCGEN	USD	SX5E	0.8938	49	0.8373	69
LLOYDS	GBP	UKX	0.7430	41	0.6631	76
	USD	UKX	0.7457	36	0.6631	76
CS	USD	SX5E	0.5505	20	0.5751	55
	CHF	SX5E	0.6427	21	0.5751	55

price. As such this regression model assumes the market CoCo price is significantly influenced by the credit spread. This is a clear discrepancy with the model for the CoCo price based on the Greeks. Notice that the Greeks are derived from the EDA pricing model which approaches the CoCo from an equity perspective.

Table 3.4 Parameter Estimates (standard error) of regression model (1) (Eq. 3.13) and Rsquared value. (*: p-value < 0.05; **: p-value < 0.01; ***: p-value < 0.001)

CoCo Name	$\hat{\alpha_1}$	$\hat{\alpha_2}$	$\hat{\alpha_3}$	$\hat{\alpha_4}$	Rsq (%)
SOCGEN 8 1/4	3.04(1.64)	5.92(36.0284)	−520.65(84.16)***	45.93(32.48)	39
SOCGEN 7 7/8	4.86(2.11)*	11.67(46.20)	−687.35(107.92)***	29.64(41.66)	41
LLOYDS 7 7/8	2.58(1.29)*	5.09(32.63)	−737.54(74.06)***	34.92(37.25)	43
LLOYDS 7 1/2	2.73(1.18)*	17.25(31.45)	−703.32(77.90)***	60.04(31.37)	43

Regression Models	
Pricing Date	30-Nov-2015
Historical Dates	01-Dec-2014 –27-Nov-2015
Number of Observations	260
Regressor	ΔP
Explanatory variables	R_{ES50} R^2_{ES50} ΔiTraxx ΔIR

In previous sections three different models are developed to explain the change in the CoCo price quote based on historical market information. These models can be summarised as:

- Model based on the Greeks without the beta coefficients (see Formula 3.9). This model predicts the performance based on the return in the underlying stock, the change in the underlying credit spread and the change in the interest rate;
- Model based on the Greeks and the beta coefficients (see Formula 3.9). This model predicts the performance based on the return in the equity index (ES50), the change in the iTraxx and the change in the interest rate;
- Regression model with explanatory variables the return in the equity index (ES50), the change in the iTraxx and the change in the interest rate (see Formula (3.13)).

In this section we evaluate each model in predicting the moves in the CoCo. The model is calibrated to one year historical data. The estimated change in CoCo price is compared with the market price change on the next day. In the following step, the window is shifted forward by one day, to find a prediction on the next day. This procedure is rolled out over time. As such each model is fitted to the data on a rolling window of one year. The historical data used to find the model is typically one year but is taken at least 2 months (or 40 observations) if older data is unavailable.

In the goodness-of-fit test, two error measures will denote the goodness of each of the models. These are the absolute error measure and the relative error measure defined by:

$$\epsilon_{ABS} = |\Delta P - \widehat{\Delta P}|$$

$$\epsilon_{REL} = \frac{\widehat{\Delta P}}{\Delta P}$$

with ΔP the observed change in CoCo price quote on the market and $\widehat{\Delta P}$ the predicted change by the model. A high value of the absolute error ϵ_{ABS} denotes that the model is bad performing in estimating the change in the CoCo price on the next day. The relative error ϵ_{REL} states if the model is denoting a right direction of the move in a sense of a down- or upward price move. In case the relative error has a positive sign, the model is predicting the move in the right direction. Although this is a useful measure in sense of directions, the measure explodes in case the market CoCo price is almost constant and does not change to the next day ($\Delta P \approx 0$).

Case study 1: SOCGEN 8 1/4 09/29/49 CoCo

The SOCGEN 8 1/4 09/29/49 CoCo market price follows a quite analogue path as the underlying spot from October 5, 2015 until October 16, 2015 although other drivers will influence the CoCo price as well (see Fig. 3.3).

On October 5, 2015 the models all overestimate the change in the CoCo quote. The market CoCo quote increased by +0.43 compared to the previous day. The model based on the Greeks and beta coefficients predicts a change of +1.74. The model based on the Greeks predicts a slightly smaller change of +1.72. The regression model is more in line with the CoCo market, predicting a change of +0.55.

The observation on October 8, 2015 is different. On this day we observed a drop in the underlying stock price around –0.95% and a daily return of the Eurostoxx 50 of –0.40%. This and other market information resulted in a drop on the CoCo of size –0.15. The models based on the Greeks predicted a change of –0.10 and –0.80. The prediction with the regression model is –0.004. The absolute error measure hence declares the model based on the Greeks with beta coefficient is fitting best the market on October 8, 2015.

The error measures are displayed in Fig. 3.4. The relative error measure shows that the regression model and the model based on the Greeks with beta coefficient is predicting a wrong direction for the CoCo price on October 12, 2015.

Case study 2: DLNA 4 3/8 06/29/49 CoCo

In previous example, the regression model is most of the days better fitting the market. But it seems that on days of extreme unforeseen events, the Greeks model will perform better. In December 2015 extreme stock price market moves were observed for Delta Lloyd bank. Delta Lloyd posted a loss of €533 mn in the first half year of 2015 due to higher-than-expected sensitivity to market volatility. The decline

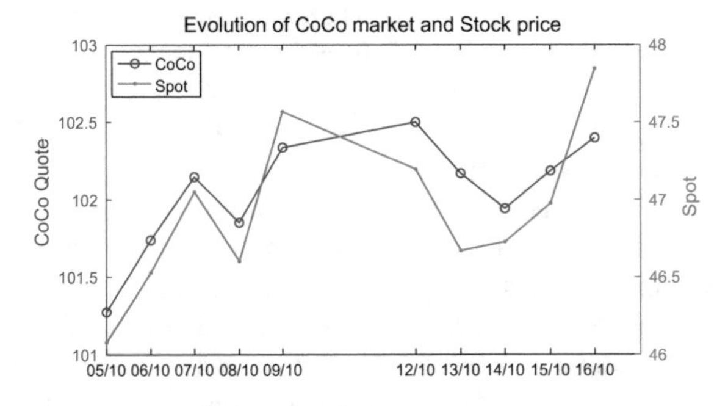

Fig. 3.3 Market SOCGEN 8 1/4 09/29/49 CoCo price quote versus the underlying stock price performance from October 5, 2015 until October 16, 2015

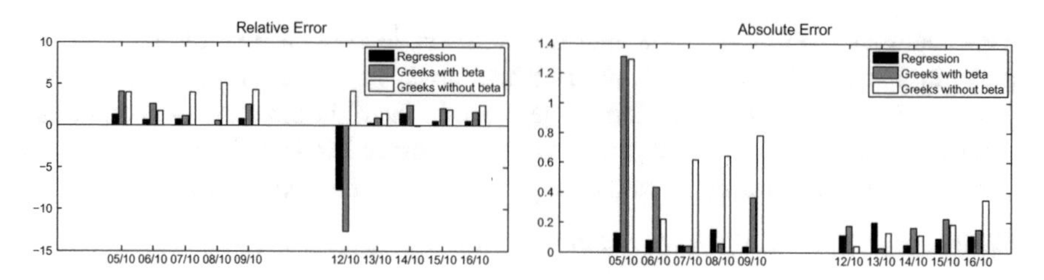

Fig. 3.4 Model performances for SOCGEN 8 1/4 09/29/49 CoCo

in asset values, stemming from increased interest rates in the second quarter, was only partly offset by a decline in the value of liabilities. Market volatility hurt Delta Lloyd's economic capital ratio, which fell below its target range. Delta Lloyds stock market experienced a huge drop by the end of August 2015. Since this issuer had no CoCo bonds outstanding, a bond of this issuer with similar features as a CoCo will be observed. We investigate the DLNA 4 3/8 06/29/49 bond price sensitivity as if it was a contingent convertible bond.

In general the model performance is the same as in previous example of one of Societe General CoCos. For example during the same period as in the previous example, the standard regression model does seem to fit best the market price moves (Fig. 3.5).

On August 24, 2015, Delta Lloyds share price falls most ever as loss raises capital concerns. The stock market of Delta Lloyd had a return of –9% over the weekend as shown in Fig. 3.5a and b. In Fig. 3.5 we observe that the Greeks model does now fit this extreme drop quite well compared to the regression model. The regression model does only denote a small drop around 0.5 units whereas the model based on the Greeks and the betas estimates the loss by –3.10. The actual CoCo market dropped by 2.39 units. On August 27, 2015 Delta Lloyds stock market recovers fast with an increase of +2.80%. This extreme impact on the CoCo market is again better fitted by the Greeks model with the beta coefficients using the Eurostoxx 50, iTraxx index and the interest rates as inputs.

Case study 3: POPSM 11 1/2 10/29/49 CoCo

Previous examples seem to indicate that the regression model is in general better fitting the CoCo market but on days of extreme unforeseen events for the specific CoCo issuer, the Greeks model will outperform the regression model. As such we also investigate the CoCo price of Banco Popular right before the trigger event took place in June 2017. In Fig. 3.6 the value of delta increases steeply near the time of trigger. Gamma becomes more negative in the neighbourhood of the trigger event. Although on May 19th, 2017 the value of gamma starts to increase. Hence the second order effect of the stock price decreases again. The impact of credit spread and interest rate on the CoCo price decreases in general near the trigger.

On May 31, 2017 Banco Popular share price experiences an extreme drop due to the announcement of Reuters. Reuters reports that the chair of the Single Resolution Board (SRB) has warned EU officials that Banco Popular "may need to be wound

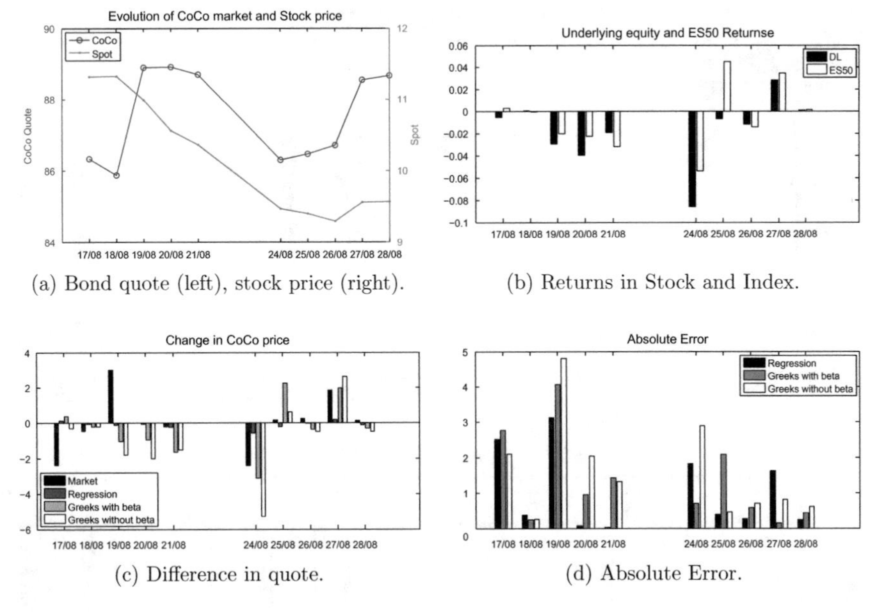

(a) Bond quote (left), stock price (right).

(b) Returns in Stock and Index.

(c) Difference in quote.

(d) Absolute Error.

Fig. 3.5 Model performances for DLNA 4 3/8 06/29/49

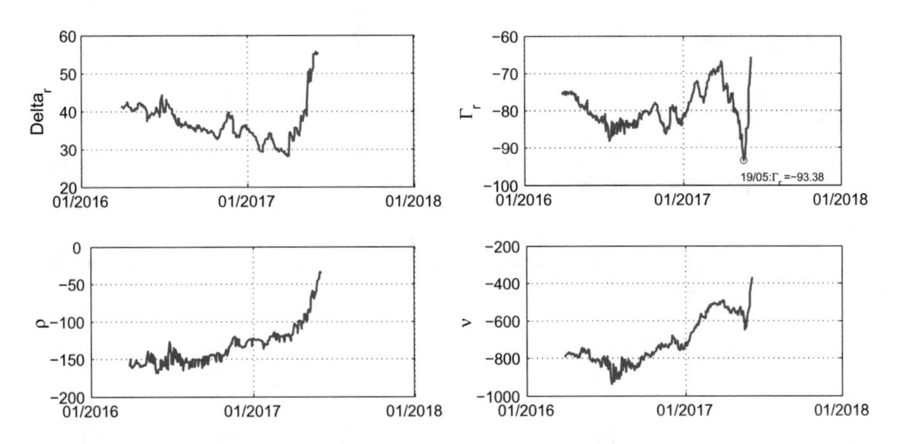

Fig. 3.6 Behaviour of the Greeks of POPSM 11 1/2 10/29/49 CoCo near the trigger event

down if it fails to find a buyer" (Guarascio 2018). This move in the underlying market price, hit the CoCo bond quote as well, with a decrease of –6.86 units. The only model able to capture a sudden more extreme drop, was the model based on the Greeks and the underlying stock price market with a prediction of –2.90. The regression model and the model of the Greeks with input the changes in the Eurostoxx 50 and iTraxx only estimate a change of resp. –0.12 and –0.20. Also on the final date of Banco Popular, the change in the CoCo price quote was best fitted by the model based on the Greeks without taken into account the beta coefficients (see Fig. 3.7). In general we observe that the sensitivity of the model based on the Greeks related to the moves

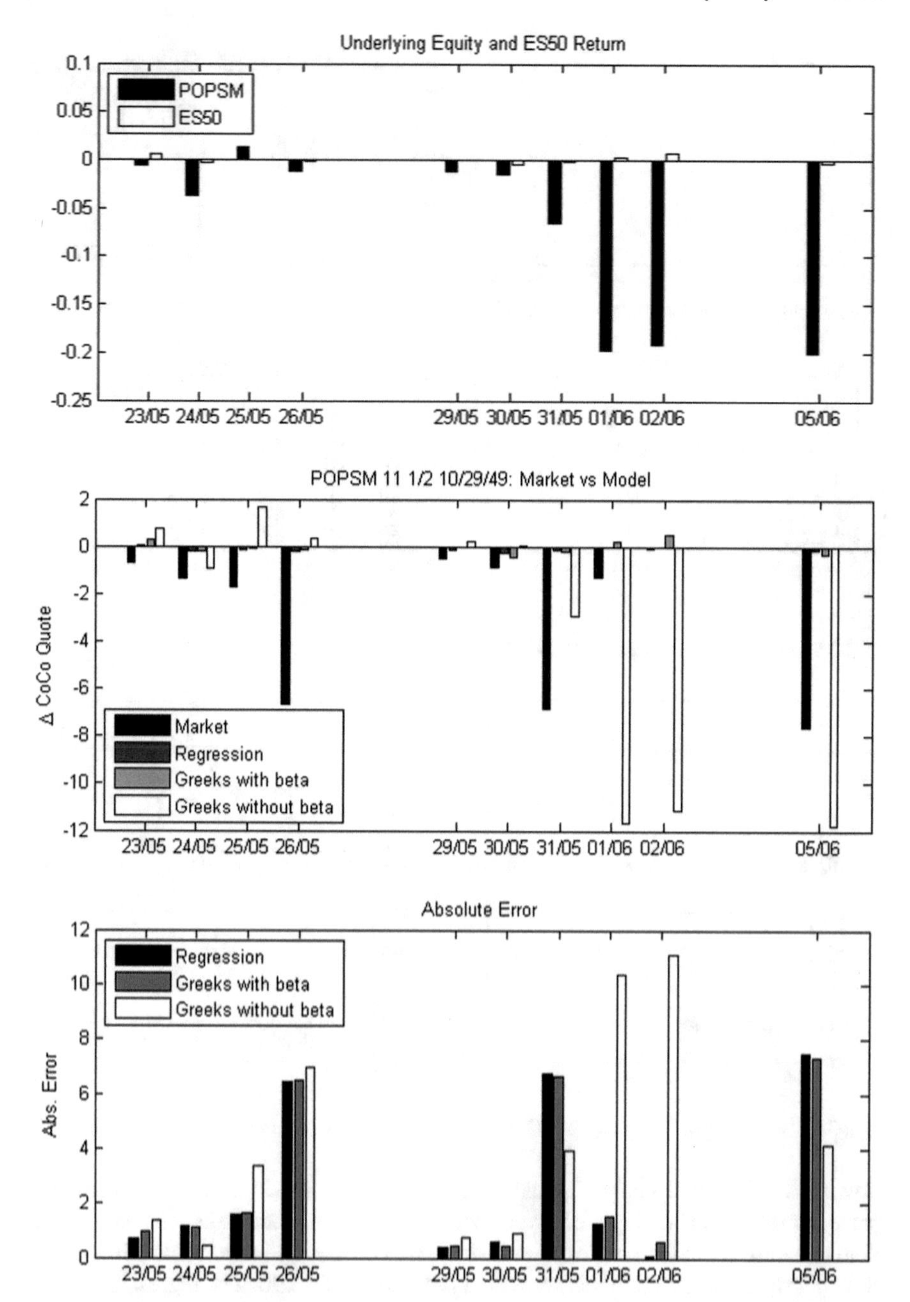

Fig. 3.7 Model performances from May 23th, 2017 until June 5th, 2017 for POPSM 11 1/2 10/29/49

in the underlying stock prices, is in line with the CoCo market price moves in times of crisis. However sometimes the model lacks information. For example on May 26, 2017 the CoCo market price jumps without significant moves in the model CoCo prices.

3.5 Conclusion

The sensitivity analysis of the CoCo price resulted in first place in estimates for the Greeks. The Greeks are partial derivatives and can be approximated with a finite difference approach using the EDA pricing model. These values can provide the CoCo investor with insides to hedge from adverse changes in the market conditions.

The Greeks determined in this chapter are expressing the sensitivity of the CoCo price to the specific underlying equity and credit market. The hedging strategy corresponding with these Greeks implies investments in the underlying equity and credit market of the specific CoCo issuer. This requires different investments for each different CoCo issuer in the portfolio. The sensitivity with respect to specific underlying markets can be translated towards a more general sensitivity with respect to some overall market indices with the beta coefficients. The hedging strategy taken into account the beta values will reduce the costs and decrease the follow-up efforts of the hedging strategy for a CoCo portfolio.

Three models for estimating the CoCo price change were evaluated in case studies. The first model was based on the Taylor expansion that combines the Greeks in order to estimate the impact on the CoCo price given changes in the underlying equity, credit spread and interest rates. In a second model, we included the beta coefficients to the model based on the Greeks. The third model is a simple regression model with the equity index, credit index and interest rate as categorical variables. In a case study of a CoCo issued by Societe General, the standard regression model did perform much better on most days than the models based on the Greeks. The model based on the Greeks seems to overestimate the impact due to a change in equity index. This overestimation is even worse when looking at the returns of the specific underlying stock, i.e. without applying the beta coefficient. With the case study of the Banco Popular CoCo, the models were compared in a stress event for the issuer. In a worse case scenarios for the specific CoCo issuer, the Greeks model seems to perform much better than the regression model.

The simple regression model indicated the credit spread as a highly significant variable in modeling CoCo price moves. Whereas the EDA pricing formula used in the models based on the Greeks, looks at the CoCo price from an equity perspective. Based on the case study results, we can point out the importance of the credit risk in non-stress situations and the equity risk in a stress situation.

Chapter 4
Impact of Skewness on the Price of a CoCo

In Chap. 2 the pricing of CoCo notes has been worked out in a market implied Black–Scholes context. However the Black–Scholes stock price model is based on several assumptions that are not related with real financial markets such as normal distributed log-returns with a constant volatility. Under the Black–Scholes model this volatility parameter is assumed to be constant for all vanilla options independent of their maturity dates or strike price. However if we use the Black–Scholes option pricing model, we can compute the implied volatility parameter of the underlying such that the model price does correspond with the market price. On the financial markets we observe different implied volatilities across the various strikes or maturities (see Fig. 4.1). This skewed pattern is referred to as the volatility smile or skew.

Multiple new models have been proposed as an effort to relax the restrictive assumptions in modeling the stock price market. In this chapter we move away from the constant volatility assumption and study CoCos in a stochastic volatility context. The Heston model (Heston 1993) is put at work to introduce uncertainty in the behaviour of volatility. The existence of a semi closed-form formula for European options pricing under the Heston model allows for a fast calibration of the model. In our approach we combined market quotes of listed option prices with credit default swaps (CDS) data.

First, we will introduce in detail the Heston model and value CoCo bonds using the CDA and EDA pricing method. In the second part of this chapter we focus on an analysis of the theoretical CoCo price with respect to the skewness in the implied volatility surface. The theory is illustrated by a detailed study of the Tier 2 (T2) CoCo with a $7\frac{5}{8}\%$ coupon issued by Barclays in 2012 and maturing on November 21, 2022. More information can be found in De Spiegeleer et al. (2017b).

J. De Spiegeleer et al., *The Risk Management of Contingent Convertible (CoCo) Bonds*, SpringerBriefs in Finance, https://doi.org/10.1007/978-3-030-01824-5_4

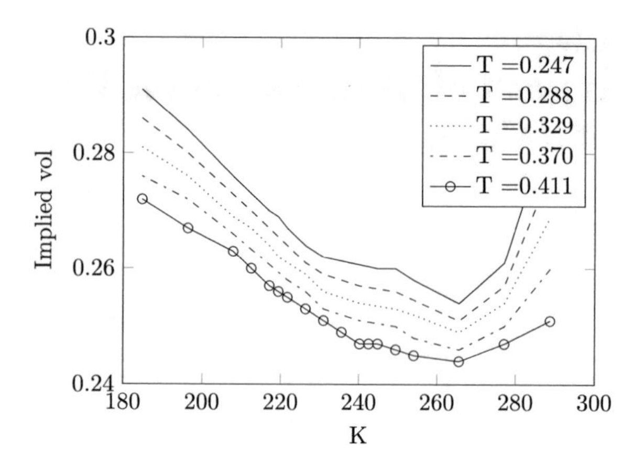

Fig. 4.1 Implied volatilities of vanilla option prices depend on strike level K and maturity T(in years)

4.1 Heston Model

Until now the Black–Scholes model was used in the CoCo valuation. It gave a first insight in the pricing and dynamics of CoCos. However, this model has significant drawbacks such as a constant volatility and the underestimation of tail risk. In reality, volatility changes with the strike price and the maturity, resulting in the so called volatility-smile. The Heston model (Heston 1993) will now be put at work as a more adequate alternative to the Black–Scholes model. The Heston model dates back to 1993 and is popular among market practitioners since it incorporates stochastic volatility in an tractable and intuitive way.

The Heston model incorporates stochastic volatility and more realistic fat-tail distributions of stock returns. Heston follows the Black–Scholes stock price model but includes the volatility as behaving stochastically over time. The (squared) volatility is following the classical Cox-Ingersoll-Ross (CIR) process (Schoutens et al. 2004). Remark that the Heston model does as such not include stock price jumps. It describes the evolution of the stock price in terms of five parameters (i.e. σ_0, ρ, λ, κ and η) with the following equations:

$$dS_t = (r - q)S_t dt + \sqrt{v_t} S_t dW_t^{(1)},\ S_0 > 0 \tag{4.1}$$

$$dv_t = \kappa(\eta - v_t)dt + \lambda\sqrt{v_t} dW_t^{(2)},\ v_0 = \sigma_0^2 > 0 \tag{4.2}$$

with S_t the stock price process, v_t the variance process. The correlated standard Brownian motions are defined by $W^{(1)} = \{W_t^{(1)}, t > 0\}$ and $W^{(2)} = \{W_t^{(2)}, t > 0\}$. A standard Brownian motion is a stochastic process $W = \{W_t, t > 0\}$ such that

1. $W_0 = 0$ almost surely.
2. W has independent increments.
3. W has stationary increments.
4. $W_{t+s} - W_t$ is normally distributed with mean 0 and variance $s > 0$: $W_{t+s} - W_t \sim N(0, s)$

Define $\Delta W_t = W_{t+\Delta t} - W_t$ which is normal distributed with zero mean and variance Δt. If $\Delta t \to 0$, ΔW_t converges to the stochastic process dW_t with variance dt.

In the Heston model, the standard Brownian motions $W^{(1)}$ and $W^{(2)}$ are correlated with $\text{Cov}[dW_t^{(1)} dW_t^{(2)}] = \rho dt$. The correlation coefficient ρ governs joint movements in the stock price and its variance. Typically, if the stock price drops, the volatility increases and ρ will receive a negative value. The more negative ρ is, the steeper the skew becomes. Moreover,

$$\lambda > 0 - \text{ is the volatility of variance}$$
$$\kappa > 0 - \text{ is the speed of mean reversion}$$
$$\eta > 0 - \text{ is the long-run mean of the variance}$$

One says that in case the Feller condition (i.e. $2\kappa\eta \leq \lambda^2$) is satisfied, theoretically the variance process never hits zero (Albrecher et al. 2007).

4.1.1 *Pricing of Vanilla Options*

Under the Heston model, the price of vanilla options can be found based on the theory explained in Carr and Madan (1999). Carr and Madan have found a semi-closed form solution to obtain the vanilla option prices numerically with the use of Fourier transforms. The derivation of this method can be found in Carr and Madan (1999).

The Carr–Madan formula is applicable for multiple stock price models. The method uses as input only the characteristic function of the logarithm of the stock price. This characteristic function has a simple equation for a lot of models, including the Heston model. The characteristic function for the Heston model ($\phi(u, t)$) as described in Schoutens et al. (2004) is

$$\begin{aligned}\phi(u,t) = {} & \exp(iu(\log S_0 + (r-q)t)) \\ & \times \exp(\eta\kappa\lambda^{-2}((\kappa - \rho\lambda iu - d)t - 2\log((1 - g_2 e^{-dt})/(1 - g_2)))) \\ & \times \exp(\sigma_0^2\lambda^{-2}(\kappa - \rho\lambda iu - d)(1 - e^{-dt})/(1 - g_2 e^{-dt}))\end{aligned} \tag{4.3}$$

with

$$d = \sqrt{(\rho\lambda ui - \kappa)^2 + \lambda^2(iu + u^2)} \tag{4.4}$$

$$g_2 = (\kappa - \rho\lambda iu - d)/(\kappa - \rho\lambda iu + d) \tag{4.5}$$

In the literature there are two formulas for the Heston characteristic function. They originate from the curve in the complex plane were the logarithmic function behaves discontinuously. This curve is typically called a branch cut. Since the Carr–Madan formula is defined and operating in the complex plane, these branching issues might encompass numerical issues. However it is proven in Albrecher et al. (2007) that no error appears when using the characteristic function of Eq. 4.3.

4.1.2 *Pricing of Exotic Options*

The EDA CoCo pricing model (see Eq. 2.13) requires the prices of barrier-type options. These options depend on the stock price path during the life-time of the option. Due to this heavy path dependence there does not exist an explicit formula for the barrier-type options in the Heston model. As such we are also obliged to rely on numerical methods to price a CoCo bond in this context.

The prices of the exotic options can be found with Monte-Carlo simulations. This will come at the price of less accurate results and higher computation times. In a Monte-Carlo method, stock price paths are generated in order to price barrier-type options. For each simulated path, the payoff of the option is calculated. In the end, the mean of these discounted payoffs lead to an approximation for the price of the barrier-type option. If more paths are simulated, the estimated price approximates the true value of this exotic option better. The method is based on the risk-neutral pricing valuation principle which states that the arbitrage free price of an (exotic) option is equal to the discounted expectation of its pay-off (Schoutens 2008).

Discrete versions of the stochastic differential equations (4.1) and (4.2) of the Heston model are required to simulate stock price paths under the Heston model. First the stochastic differential equations are expressed in the following integral form:

$$\begin{aligned} S_{t+dt} &= S_t + \int_t^{t+dt} (r-q)S_t dt + \int_t^{t+dt} \sqrt{v_t} S_t dW_t^{(1)} \\ v_{t+dt} &= v_t + \int_t^{t+dt} \kappa(\eta - v_t) dt + \int_t^{t+dt} \lambda \sqrt{v_t} dW_t^{(2)} \end{aligned}$$

Secondly, these integrals will be approximated using Îto calculus. The Euler scheme is given by:

$$S_{t+\Delta t} = S_t(1 + (r-q)\Delta t + \sqrt{v_t}\sqrt{\Delta t}\epsilon_1) \tag{4.6}$$

$$v_{t+\Delta t} = v_t + \kappa(\eta - v_t)\Delta t + \lambda\sqrt{v_t}\sqrt{\Delta t}\epsilon_2 \tag{4.7}$$

with ϵ_1 and ϵ_2 correlated standard normal random numbers with correlation coefficient ρ. To obtain the correlated random numbers, first two independent standard normal random variables are generated, ϵ and $\epsilon^{\star}$. Afterwards define:

$$\begin{aligned} \epsilon_1 &= \epsilon \\ \epsilon_2 &= \rho\epsilon + \sqrt{1-\rho^2}\epsilon^{\star} \end{aligned}$$

The Milstein scheme is another numerical approach and is defined by:

$$S_{t+\Delta t} = S_t(1 + (r - q)\Delta t + \sqrt{v_t}\sqrt{\Delta t}\epsilon_1) \tag{4.8}$$

$$v_{t+\Delta t} = v_t + (\kappa(\eta - v_t) - \lambda^2/4)\Delta t + \lambda\sqrt{v_t}\sqrt{\Delta t}\epsilon_2 + \lambda^2\Delta t(\epsilon_2)^2/4 \tag{4.9}$$

Under these approximations, the variance can become negative even under the Feller condition. In order to solve this problem a lot of heuristic rules exist but none is without including extra errors in the truncation. In the first solution, called absorption, the variance $v_{t+\Delta t}$ is set equal to zero in case it becomes negative. Another possibility is the reflection method where one takes absolute values for v_t in Eqs. 4.7 and 4.9. A last method which is applied here, is truncation. We can use full or partial truncation. In partial truncation only the variance v_t under the square root in Eqs. 4.7 and 4.9 is replaced by $\max(0, v_t)$. In the example below full truncation is used where v_t in the second term (the drift) and the third term (the diffusion of the variance) of these equations is replaced by $\max(0, v_t)$.

4.1.3 Calibration

The goal of the calibration is to select the set of Heston parameters such that the model price is as close as possible to the real prices observed in the market. There exist multiple metrics to measure the error between the two prices. For example, the weighted Root Mean Squared Error (w-RMSE) and the Average Percentage Error (APE) defined by:

$$\text{w-RMSE} = \sqrt{\sum_i w_i(P_i - \hat{P}_i)^2} \tag{4.10}$$

$$\text{APE} = \sum_i \frac{|P_i - \hat{P}_i|}{\hat{P}_i} \tag{4.11}$$

with P_i the market price, $\hat{P}_i$ the model price and w_i the corresponding weight. The sum of all weights w_i is often taken equal to one. In the calibration procedure applied below we will minimize the weighted RMSE error measure.

The parameters are in most cases calibrated on plain-vanilla option price data. However, the available market put and call prices have in general a maturity around one to three years and a strike in the neighbourhood of the current stock price. Deep out-the-money options with long maturity are extrapolated almost all the time. Most CoCos have on the contrary a maturity around five years or longer with a conversion in case of default. Since the parameters will be applied to price a CoCo, we would like to include data with a higher maturity and a smaller strike in the calibration. Credit Default Swap (CDS) data appears as a solution.

A CDS is a bilateral contract that will protect the CDS buyer from a default on a debt. In return for this protection, the buyer of the CDS pays the coupon leg. These payments can be made on for example quarterly instalments but it is also possible to pay them as a lump sum, called the upfront cost, at the beginning of the contract. After a default, the regularly made payments for protection stop and the buyer of the CDS receives the par value minus the recovery value from the seller.

In an heuristic way, the payoff of a zero-recovery CDS in case of default equals the payoff of a binary put option. The default can be interpreted as the stock falling below a low strike. The upfront cost of a zero-recovery CDS then can be considered equal to the price of such a binary put option. The strike of the put option is taken equal to 15% (De Spiegeleer et al. 2014). As such a market price for a deep-out-the-money digital put option with a high maturity can be derived using a CDS dataset. Remark that there is a small mismatch as to the time the payments take place. The digital put option pays out at maturity in case the stock ends below the strike, whereas the CDS pays out at the time of default.

There are only a few CDS prices available in the market. We will combine the vanilla option market prices together with CDS prices in the calibration. In order to fit these CDS data best higher weights are attributed to the CDS data in the weighted-RMSE.

4.2 Case Study - Barclays

We illustrate the methodology for a Barclays CoCo (Table 4.1) on March 27, 2014. Barclays had a stock price of 230.95 GBp on this day. The market data used in the case study consist of the list of vanilla option prices, CDS prices and interest rates.

Since the CoCo market price is given on the market, we will search for the implied trigger level $\widehat{S^\star}$ for which the model price is equal to the market price. This is the stock level which is assumed to be hit for the first time when this CoCo is written down. The calculations are explained in four steps going from the CDS and vanilla options to the implied barrier of the CoCo.

STEP 1: Collection of Market Data

The calibration is based on market data of vanilla options and CDS spreads of Barclays. The recovery rate of the senior CDS is assumed to be 40% (Hickman 1958; Altman and Kishore (1996)). We have assumed a constant dividend yield of 2.4%.

The calibration uses out-the-money European call and put options. The given datapoints, i.e. the put and call prices, are presented with respect to their strike and maturity in Fig. 4.2. The stars denote the put options derived from the CDS data. The strike of these options was assumed to be 15% of spot. The maturities of the CDS data go from one to ten years. The other data points have a strike in the neighbourhood of the current stock price and a maturity below two years. A reasonable CoCo's implied volatility would be situated in the upper left corner with a low strike and high maturity.

Table 4.1 CoCo bond issued by Barclays

ISIN	US06740L8C27
Price	110.378
CET trigger	7.00
Coupon	7.625
Frequency	Semi-annual
Maturity	21/11/2022
Face value	1000
Currency	USD
Issue date	21/11/2012
Issue size	(bn) 312
Callable	No
Coupon cancellation	No
Regulatory treatment	TIER 2
Loss absorption	Full write down

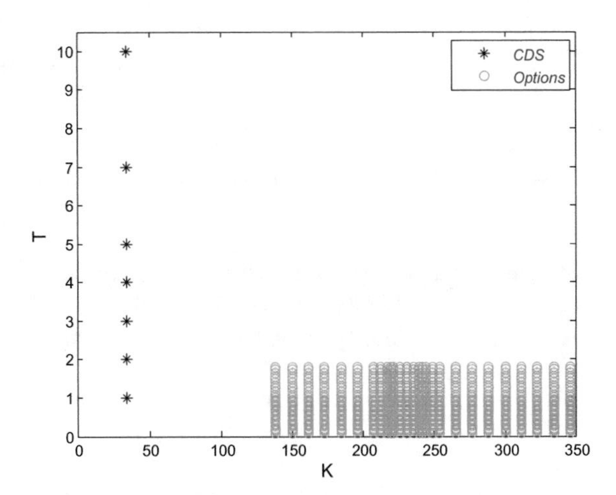

Fig. 4.2 Available data from CDS prices (star) and vanilla options (dot) used in the calibration represented with strike and maturity

STEP 2: Calibration

In the calibration step we select the parameter values where the weighted RMSE error is minimized. A higher weight is given to observations that are more important in the fitting. These observations are closely located near the area of CoCo bonds i.e. a high maturity and low strike level.

The following weighting scheme is applied to the observations. We divide the data in two sets: the 'CDS data group' (i.e. the stars in Fig. 4.2) and the 'equity data group' (i.e. the dots in Fig. 4.2). First, he total weight is equally divided over these two groups. The CDS data group and the equity data group receive each 50%. Second, the weight of a group is divided over each individual observation. Each option in

Table 4.2 Starting values and optimal parameters from the calibrated Heston model

	ρ	λ	κ	η	σ_0
Starting	–0.5	0.2	0.4	0.2	0.5
Optimal	–0.54	0.19	0.02	0.90	0.24

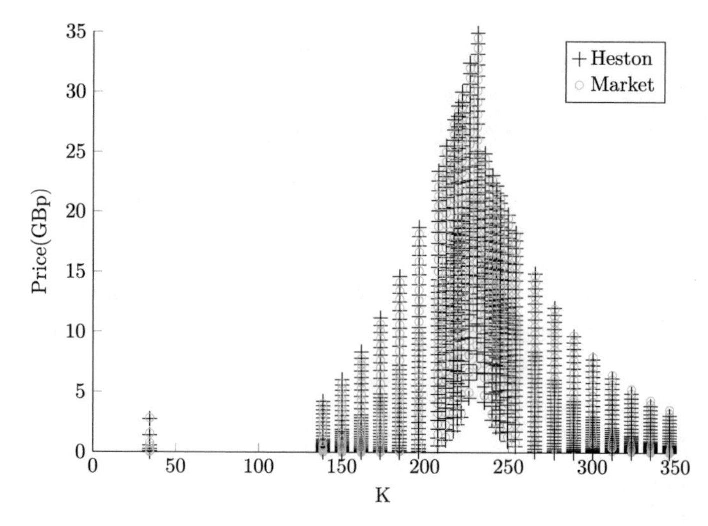

Fig. 4.3 Calibration of Heston with the market prices (dot) and the model prices (plus)

the equity data group is equally important. For the CDS data group, the higher the maturity of the CDS, the more weight it will receive. We assign the observation with the ith smallest maturity in the CDS data a weight w_i of

$$w_i = \frac{1}{2} \frac{2^i}{\sum_k 2^k} \tag{4.12}$$

The minimization of the weighted RMSE is performed using the Nelder-Mead algorithm (Nelder and Mead 1965) which is a commonly used non-linear optimization technique. The five optimal parameters obtained from the calibration are given in Table 4.2. The graphical presentation of the optimisation results is given in Fig. 4.3 where the model prices of the options are shown together with the market prices. The weighted RMSE with these optimal parameters is 0.27 and the APE is 0.03.

In Table 4.3 the European put prices derived from the CDS data are shown for the Heston model. The corresponding market data are also given. If the maturity increases, the error made in the estimation decreases. There is still place for improvement in the CDS prices with smaller maturities.

The same procedure can be applied to the Black–Scholes model and results in an optimal implied volatility parameter of 16.95%. The weighted RMSE and APE are resp. 20.55 and 0.83. It is clear that Black–Scholes model will not fit the data as

Table 4.3 Put option prices (GBp) of the CDS data obtained from the market and the Heston model

T	1	2	3	4	5	7	10
Market	0.024	0.088	0.236	0.480	0.753	1.540	2.814
Model	0.010	0.010	0.077	0.261	0.561	1.389	2.735

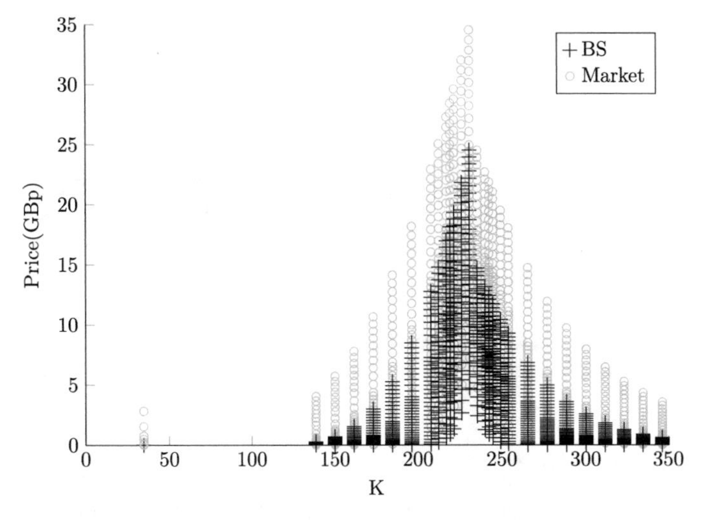

Fig. 4.4 Calibration of Black–Scholes model with the market prices (dot) and the model prices (plus)

well as the Heston model. Notice that the Black–Scholes model only includes one volatility parameter. As such the fitting of the model is less accurate compared to the five parameters Heston model. The Black–Scholes model underestimates a lot of option prices (see Fig. 4.4).

In a second approach to estimate the Black–Scholes volatility parameter we will only use the CDS data. Based on a linear interpolation of the implied volatilities of the CDS data we find an implied volatility of 42.50%. The model prices and market prices are shown in Fig. 4.5. This higher implied volatility leads to overestimation of the market prices for the vanilla options. But it is clear that the model fits the CDS data better. The weighted RMSE is 24.49 and APE is 0.75. Due to a sufficiently accurate estimation for the CDS data, we will use this volatility parameter of 42.50% for the Black–Scholes model in the next steps of our derivation to obtain the implied barrier.

STEP 3: Implied Barrier of CoCo

Once the model parameters are derived, we calculate the CoCo prices for a range of trigger levels $S^{\star}$ as explained in previous chapter. Both the EDA (Eq. 2.13) and CDA (Eq. 2.10) approach can be used to find a corresponding CoCo price.

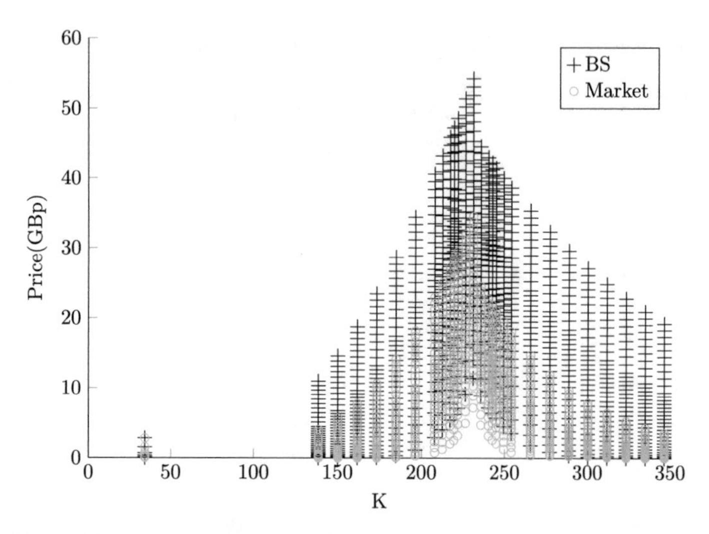

Fig. 4.5 Black–Scholes model prices (plus) with the volatility parameter estimated from a linear interpolation of the implied volatilities of the CDS data compared with the market prices (dot)

Equity Derivatives Approach

The path-dependent options in the EDA pricing method can be derived under the Heston model with a Monte-Carlo method. In this case study, stock price paths are simulated using the Milstein scheme and 5.000.000 stock paths simulated for each pricing, each consisting of 2.500 time steps covering the full maturity of the CoCo.

This sequence of CoCo prices for the Heston model can be compared with the CoCo prices derived under the Black–Scholes model (see Eq. 2.13). In Fig. 4.6 the model prices are shown in function of the barrier level. The implied barrier is the trigger level for which the model price equals the market price of the CoCo. Under the EDA CoCo pricing model, this barrier is set at 45.36 GBp or at 19.64% of the current stock for the Heston stock price model. Under the Black–Scholes model we obtain a lower implied barrier of 33.35 GBp or 14.44%.

The probability of the stock falling below the implied barrier can be derived from a similar Monte-Carlo method and equals 35.18% under the Heston model. Under the Black–Scholes model, the probability that Barclays' CoCo will get triggered is equal to 34.94%.

Credit Derivatives Approach

In the CDA CoCo pricing method the price of a CoCo can be derived from the trigger probability. Under the CDA model, this probability corresponds to the probability that the stock price drops below the barrier level S^*. From this trigger probability one can back out the CoCo spread (see Eq. 2.8). This CoCo spread together with the risk-free rate results in a yield. At last, the future cash flows (i.e. the coupon payments and the repayment of the initial value) are discounted with the corresponding yield resulting in the CoCo price as given in Eq. 2.10.

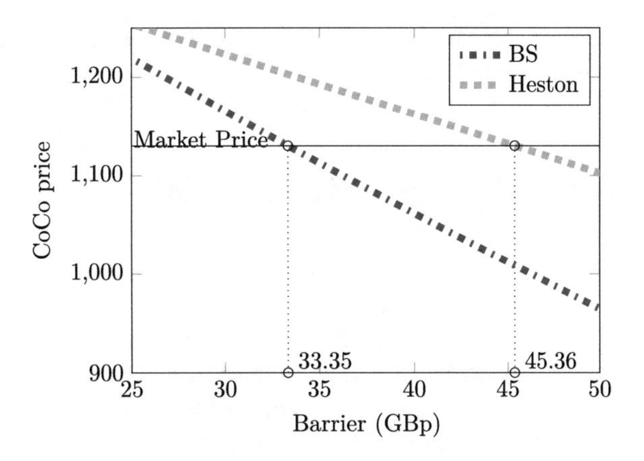

Fig. 4.6 Implied barrier of the EDA CoCo price method for Black–Scholes model and Heston model

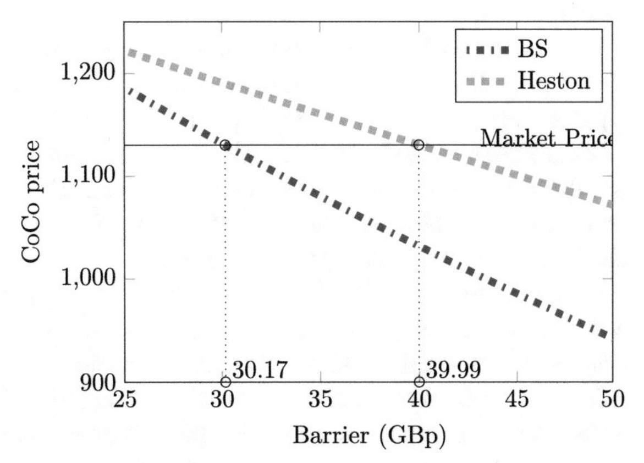

Fig. 4.7 Implied barrier of the CDA CoCo price method for Black–Scholes model and Heston model

For the Black–Scholes model, we have an explicit formula for the trigger probability. Under the Heston model, the trigger probability is again derived from Monte-Carlo simulations. For the CDA pricing method (Fig. 4.7), the implied barrier is 39.99 GBp or 17.32% of the initial stock price level under the Heston model. For the Black–Scholes model the implied barrier equals 30.17 GBp (or 13.06%). Both barrier levels do correspond with a trigger probability of 25.81%.

4.3 Sensitivity to Parameters of the Heston Model

In this section, the sensitivity of the price of a CoCo with respect to the parameters of the Heston model is investigated. We take the Barclays CoCo of the previous section as main example. We investigate the sensitivity in the current, a distressed and a non-distressed situation. The distressed setting is assuming the stock is close to the trigger level, namely 75% of spot; the non-distressed case assumes a barrier at

15% of spot. We put each of the studies in perspective with the skew of the implied volatility surface.

4.3.1 Example of Barclays' CoCo

First we discuss the impact of a change in one of the parameters of the Heston model on the skew and on the CoCo price. From a set of Heston parameters a curve of implied Black–Scholes volatilities is calculated. The observed skew is the logical consequence of looking at the Heston process through a pair of Black–Scholes glasses. The CoCo price corresponding with the same set of Heston parameters is also derived using 1.000.000 paths in the Monte-Carlo simulations and the EDA pricing method.

In Fig. 4.8, each plot in the left column corresponds to the analysis of the Barclays' CoCo price for one single parameter of the Heston model. In the right column the figures show the implied volatility for different parameter values. In each picture, one parameter of the Heston model changes in value and the changes in the volatility smile due to this one parameter can be observed. The other parameters are fixed and as defined in Table 4.2. The implied barrier, $\widehat{S^{\star}}$, is set equal to the result obtained from the case study and equals 45.36 GBp.

Each parameter of the Heston model has its impact on the skew of the implied volatilities and the CoCo price. We analyse the results for each of the five Heston parameters. Since the correlation coefficient, ρ is negative, a decrease in the stock price will result in an increase in the implied volatility. This is the so-called leverage effect. When ρ gets closer to zero, the negative relation between the stock price process and the variance process is reduced and the implied volatility curve will be more flat as in the Black–Scholes model where there is a constant volatility. When ρ decreases, the left tail volatility increases and the stock is more likely to hit the implied barrier leading to lower CoCo prices.

The parameter λ, denotes the volatility of the variance process. First the CoCo price increases when we increase λ but after a while the CoCo price starts decreasing. For λ equal to 1, the extreme left tail implied volatility is high. Therefore the stock is more likely to hit the barrier and one could argue that CoCo price should decrease. When λ decreases, this extreme left tail volatility is lower and hence the price should increase. However, there is a second effect: increasing λ decreases the ATM volatility, leading to an opposite effect.

The speed of mean reversion of the variance process is denoted by κ. If κ increases, the variance process is converging faster back to its mean variance level. Since in our case the mean reverting volatility level $\sqrt{\eta} > \sigma_0$, we are moving faster towards a high volatility level and hence a lower CoCo price. If the long-run mean level of the variance process, η, increases, it is more likely that the stock will hit the implied barrier, since on average we are in a higher volatility regime. The same reasoning holds for the initial volatility σ_0.

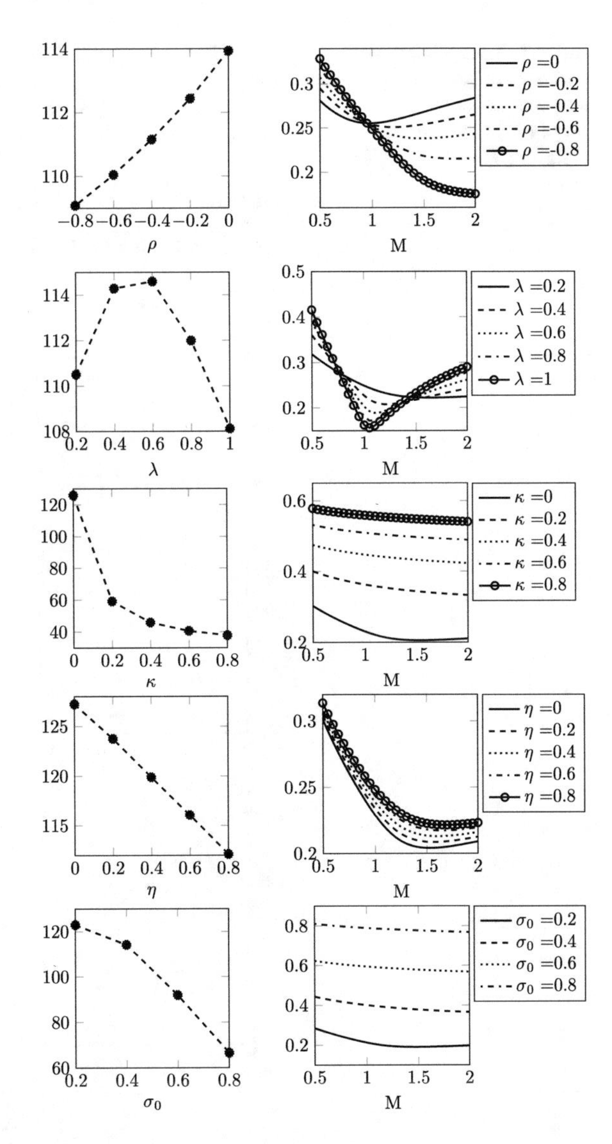

Fig. 4.8 Left: Price of Barclays' CoCo; Right: Corresponding volatility smiles

4.3.2 Distressed Versus Non-distressed Situation

The impact on the CoCo price for the parameters of the Heston model can also be compared between a situation of stress and non-stress. These situations are modeled by setting the initial stock price close to the implied barrier or far away.

The results are shown in Fig. 4.9. The left column displays a non-stress situation. We start from the Barclays setting but assume now a barrier equal to 15% of spot. Hence it is unlikely that the barrier will be hit immediately. In the right column,

Fig. 4.9 Price of Barclays' CoCo. Left: Non-stress situation: $H = 15$, $S_0 = 100$; Right: Stress situation: $H = 15$, $S_0 = 20$

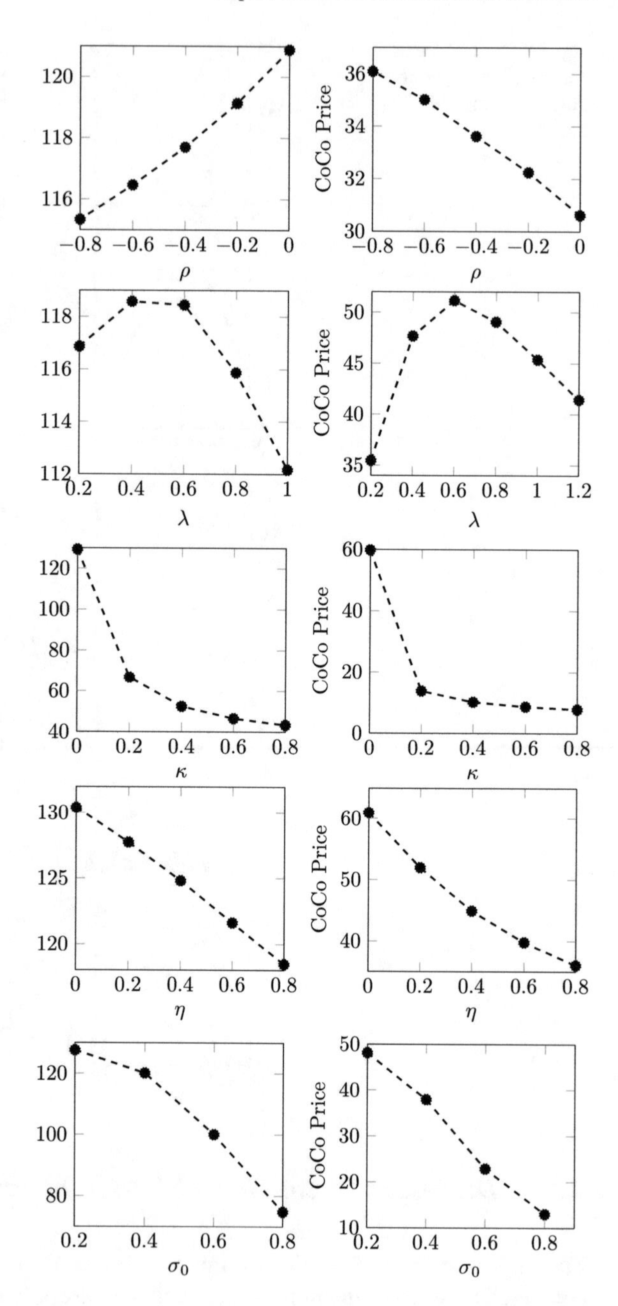

results for a distressed situation are displayed where the barrier is set at 75% of the initial stock, and hence it is likely that the barrier will be hit quite fast. Immediately, a large difference can be observed between the two situations. The relation between the CoCo price and the correlation coefficient inverts. When the stock is close to the barrier, the change of ρ has another impact on the CoCo price. An increase in ρ (less negative) will increase the price and result in a flatter volatility curve.

The impact of the other parameters is only slightly different. In times of stress, the CoCo price curve for the volatility of variance, λ, does not decrease much for λ large. The ATM implied volatility will be of more importance in determining the CoCo price compared with the left tail implied volatilities in times of stress. The curve corresponding to the variation of the mean reversion level, η becomes more convex when the stock comes closer to the barrier. The CoCo price decreases faster for an increase of κ in times of stress. The impact of the initial volatility, σ_0, remains almost the same.

Let us further investigate the relation with the correlation coefficient ρ. The positive linear relation between the CoCo price and the correlation coefficient becomes negative closer to the barrier. The turning point corresponds with an implied barrier H for which the change in ρ has no impact on the CoCos' price. An estimate for this turning point can be found as follows. Let us denote the CoCo price in function of the correlation coefficient with $P(\rho)$. An estimate of the slope of the CoCo price curve is denoted with $\delta^\star$ and can be derived by:

$$\delta^\star = \frac{\partial P}{\partial \rho} \approx \frac{P(\rho_2) - P(\rho_1)}{\rho_2 - \rho_1}$$

with $(\rho_1, O(\rho_1))$ and $(\rho_2, O(\rho_2))$ two points of this curve. Notice that $\delta^\star$ is a kind of "Greek" which measures the sensitivity of the CoCo price with respect to the parameter ρ. The implied barrier H for which the slope becomes zero is the turning point.

In Fig. 4.10 the value of $\delta^\star$ is shown in function of different values for the barrier H. The same conclusion follows for the different choices of ρ_1 and ρ_2. The relation with ρ changes direction when the barrier is around 40% of the initial stock price. The slope is positive if the implied barrier is below 40% of the initial stock price. When the implied barrier is higher than 40%, the CoCo price decreases for increasing ρ.

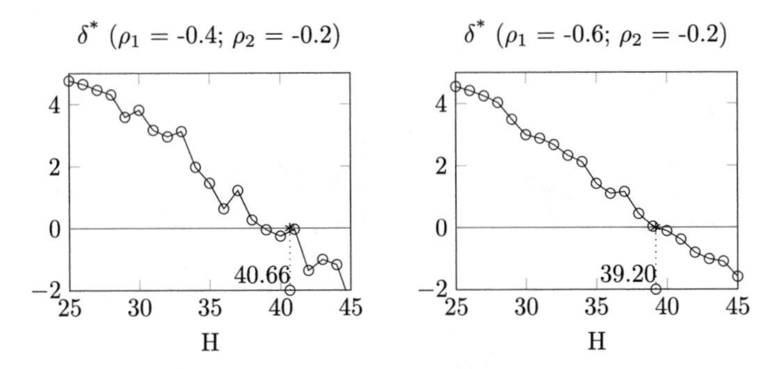

Fig. 4.10 Slope $\delta^\star$ in function of the barrier H with $S_0 = 100$

4.4 Implied Volatility Surface

Under Black–Scholes pricing, one assumes a constant volatility and a flat implied volatility surface. However under the Heston model, it is possible to implement a more market conform volatility skew in the pricing of CoCos. The relation between the CoCo price and the skew is further illustrated here for the Barclays' CoCo.

We create different sets of parameters which result in a very similar At-The-Money (ATM) implied volatility but show different skew profiles. For example in Table 4.4 each parameter set results in a similar implied volatility around 25%. However the skewness, measured by the difference in implied volatility for a 110% moneyness (i.e. $K/S_0 = 110\%$) and the implied volatility at 90% moneyness, ranges from 0 up to –2%. This results in a CoCo price change from 109 up to 114. The different skew profiles are visualised in Fig. 4.11. The CoCo prices are shown together with a 95% confidence interval defined by

$$\frac{1}{m}\sum_i P_i \pm 1.96\sqrt{\mathrm{Var}(P_i)} \tag{4.13}$$

with $\mathrm{Var}(P_i)$ the sample variance of a range of CoCo prices P_i derived from the multiple Monte-Carlo paths ($i = 1, \ldots, m$ with $m = 1.000.000$).

Table 4.4 CoCo price and skewness ($.10^{-2}$) for different parameter sets of the Heston model with constant implied volatility at 100% moneyness

Parameters (ρ, λ, κ, η, σ_0)	$\sigma_{M=100\%}$	Skew	CoCo price
(**0**, 0.1864, 0.0231, 0.8968, 0.2415)	0.2556	0.06	113.98
(**–0.2000**, 0.1864, 0.0231, 0.8968, 0.2415)	0.2537	–0.64	112.45
(**–0.4000**, 0.1864, 0.0231, 0.8968, 0.2415)	0.2518	–1.36	111.12
(–0.5378, 0.1864, 0.0231, 0.8968, 0.2415)	0.2506	–1.86	110.41
(**–0.6000**, 0.1864, 0.0231, 0.8968, 0.2415)	0.2500	–2.09	110.06
(**–0.8000**, 0.1864, 0.0231, 0.8968, 0.2415)	0.2483	–2.83	109.01

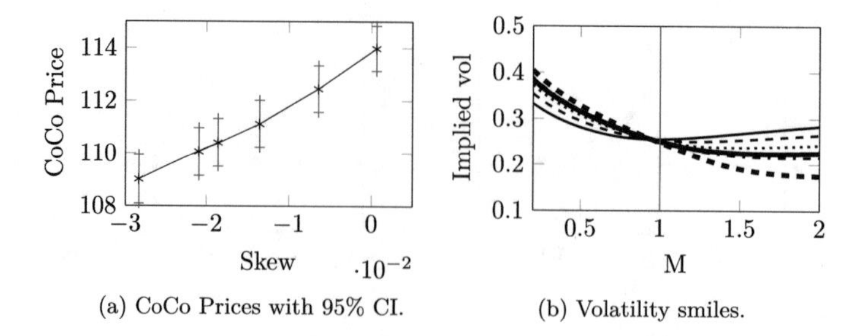

(a) CoCo Prices with 95% CI. (b) Volatility smiles.

Fig. 4.11 CoCo price sensitivity due to different skews with constant ATM implied volatility

Table 4.5 CoCo price and skewness ($.10^{-2}$) for different parameter sets of the Heston model with constant implied volatility at 20% moneyness

Parameters (ρ, λ, κ, η, σ_0)	$\sigma_{M=20\%}$	Skew	CoCo price
(**0**, 0.1864, 0.0231, 0.8968, **0.3200**)	0.3846	0.03	104.58
(**–0.2000**, 0.1864, 0.231, 0.8969, **0.2880**)	0.3841	–0.57	107.30
(**–0.4000**, 0.1864, 0.0231, 0.8968, **0.2600**)	0.3845	–1.29	109.28
(–0.5378, 0.1864, 0.0231, 0.8968, 0.2415)	0.3841	–1.86	110.40
(**–0.6000**, 0.1864, 0.0231, 0.8968, **0.2350**)	0.3849	–2.13	110.62
(**–0.8000**, 0.1864, 0.0231, 0.8968, **0.2100**)	0.3841	–3.16	111.91

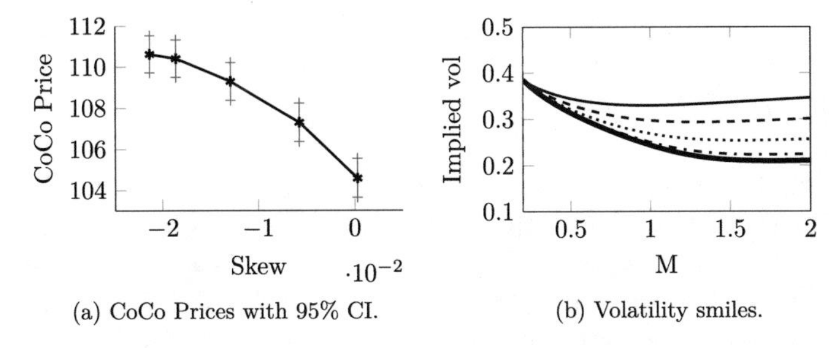

(a) CoCo Prices with 95% CI. (b) Volatility smiles.

Fig. 4.12 CoCo price sensitivity due to different skews with constant implied volatility for moneyness 20%

Table 4.6 CoCo price and skewness ($.10^{-2}$) for different parameter sets of the Heston model with constant implied volatility at 20% moneyness

Parameters (ρ, λ, κ, η, σ_0)	$\sigma_{M=20\%}$	Skew	CoCo price
(–0.5378, 0.1864, 0.0231, 0.8968, 0.2415)	0.3841	–1.86	110.36
(–0.5378, **0.2000**, 0.0231, 0.8968, **0.2310**)	0.3845	–2.07	111.52
(–0.5378, **0.2200**, 0.0231, 0.8968, **0.2150**)	0.3849	–2.39	113.15
(–0.5378, **0.2500**, 0.0231, 0.8968, **0.1900**)	0.3849	–2.88	115.56
(–0.5378, **0.3000**, 0.0231, 0.8968, **0.1480**)	0.3849	–3.45	118.88
(–0.5378, **0.3500**, 0.0231, 0.8968, **0.1070**)	0.3849	–3.41	121.14

In Table 4.5 the parameter values ρ, κ and η are fixed. The implied volatility is now kept constant at the 20% of spot level. By changing the parameter values for ρ and σ_0, we create different skewness in the volatility smiles which impacts the CoCo prices (see Fig. 4.12).

In Table 4.6 the parameter values of λ and σ are altered in order to create a different shape in the volatility smiles. The implied volatility is kept constant at the moneyness of 20%. This creates an even more significant impact on the CoCo prices ranging from 110 to 121 (see Fig. 4.13).

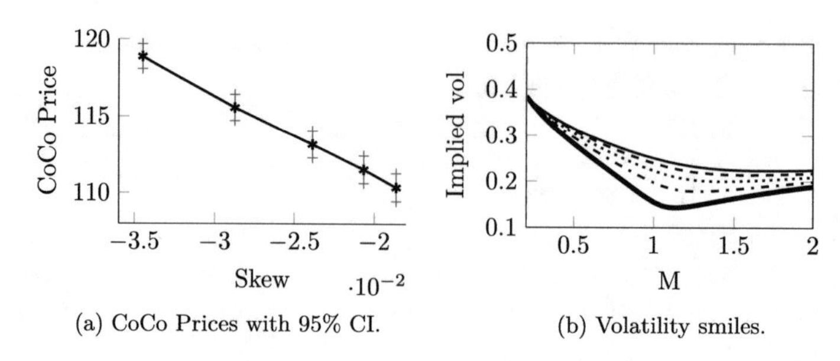

(a) CoCo Prices with 95% CI.

(b) Volatility smiles.

Fig. 4.13 CoCo price sensitivity due to different skews with constant implied volatility for moneyness 20%

4.5 Conclusions

The impact of skew on the pricing of CoCo bonds was investigated by employing a stochastic volatility model (Heston) able of capturing the market skew accurately. We operate in a market implied setting and use a derivatives approach for the pricing of the CoCo bond. We have analyzed the effects of changes in skew of the implied volatility surface of CoCo bonds in a current, a distressed and non-distressed setting. A numerical estimation of the turning point, i.e. an implied barrier level H for which the change in ρ has no impact on the CoCos' price, is around 40% of the initial stock price.

We conclude that the equity derivatives approach can serve as a good start model to derive the price of a CoCo, but users should be aware that there can be a material impact due to skew. We have priced an example CoCo under different implied volatility curves with a fixed implied volatility at the 20% of spot strike level and have observed price impacts up to 10%. We conclude CoCos are significantly skew sensitive and argue for advanced models to accurately capture related risks in the assessment of CoCos.

Chapter 5
Distance to Trigger

Some argue that the distance between the bank's current CET1 ratio and the triggering level is irrelevant given the presence of the PONV regulatory trigger. Nevertheless, many market practitioners seem to believe that these buffers of regulatory capital provide a useful tool for a relative valuation.

In this chapter we apply the implied CET1 volatility model in order to define a risk-adjusted distance to trigger.[1] In the implied CET1 volatility model the CoCo bonds are described as financial derivative instruments contingent on the CET1 level. The CET1 level is assumed to follow Black's model with only one volatility parameter. From this CoCo price model, we can derive the implied CET1 volatility. It is the volatility parameter such that the model CoCo prices match exactly with the market quote. The numerical results show how different CoCos issued by the same bank and sharing a similar CET1 trigger have almost identical implied CET1 volatility levels. The same results indicate also the difference in market risk between Tier 2 and Additional Tier 1 CoCo bonds. Hence it seems natural to adjust the distance to trigger by taking into account the implied CET1 volatility.

The ability to obtain an implied level for the CET1 volatility offers furthermore other interesting results. It allows to look at the severity of the stress test imposed by the ECB on European banks in November 2014. In that perspective, we are also able to infer an implied PONV CET1 level and an implied coupon cancellation level. More information can be found in De Spiegeleer et al. (2017a).

[1] This chapter is based on De Spiegeleer, J., S. Höcht, I. Marquet, and W. Schoutens (2017a). "CoCo bonds and implied CET1 volatility". In: Quantitative Finance 17.6, pp. 813–824. https://doi.org/10.1080/14697688.2016.1249019.

J. De Spiegeleer et al., *The Risk Management of Contingent Convertible (CoCo) Bonds*, SpringerBriefs in Finance, https://doi.org/10.1007/978-3-030-01824-5_5

5.1 Distance to Trigger Versus CoCo Spread

A quick and simple version for the distance to trigger (D) of a CoCo bond is the ratio of the Common Equity Tier 1 (CET1) Ratio under Basel III and the contractual CET1 trigger level of the CoCo bond.

$$D = \frac{\text{CET1 Ratio}}{\text{Contractual Trigger}} \tag{5.1}$$

The triggering of a bond corresponds to a situation where $D < 1$ before the maturity date.

First, we compare the CoCo spreads (cs_{CoCo}) and the distance to trigger (D) of 26 different contingent convertibles issued by 9 different banks. Only the results for write-down AT1 bonds are taken into consideration. CoCos with equity conversion and Tier 2 CoCos are at this stage left out of the scope. The yearly coupon rate and the CET1 trigger level are also shown for each CoCo bond. The CET1 ratios reported in Table 5.1 have a cut-off date at November 16, 2015. These capital ratios all reflect the third quartile results of 2015. The exact publication date of the 3Q 2015 result is also shown. The CoCo spreads are derived at these corresponding publication dates.

Linking the value of the CoCo bond solely to the distance to trigger is similar to valuing an option on the difference between the strike and the value of the underlying share. The more the option is in the money, the higher its market value. The distance to trigger can be compared with the intrinsic value of the CoCo bond. Doing so, one makes abstraction of the volatility of the CET1 ratio. For example compare the Nordea bond with an 8% CET1 contractual trigger and the Unicredito CoCos with a 5.125% trigger. They both have a very similar value for D. The values of the CoCo spreads are very different on the other hand. The NORDEA 5.5% bond has $D_{\text{Nordea}} = 2.04$ with $cs_{CoCo} = 466$bps. The UNICREDITO 6.75% contingent convertible shares almost the same value for D ($D_{\text{Unicredito}} = 2.05$) but has $cs_{CoCo} = 658$bps. The market value of the Unicredito bond incorporates a higher credit spread and tells us more on how investors look on this particular issue. Investors without any doubt regard Nordea as a much safer institution than its Italian counterpart. Its CoCo bond shares the same value for D, but market practitioners attach clearly a lower probability that D_{Nordea} drops below one before the maturity date of the bond. Therefore the concept of D should be expanded by taking into account the volatility of the CET1 ratio.

For the 26 different contingent convertibles of Table 5.1, the implied CET1 volatilities were calculated using their market CoCo price. For each of the banks, this calculation was performed immediately after the release of its capital ratio for Q3 2015. Logically, these calculation dates hence differ slightly from bank to bank. For example the values of $\overline{\sigma}_{CET1}$ for Unicredito were calculated on November 11, 2015. This is three weeks after Credit Suisse and Nordea released their results. The results of the calculation can be found in Table 5.2. An observation that stands out, is the fact that for the identical CET1 levels, the implied CET1 volatilities of a particular issuer are very similar. The CoCo bonds are fairly priced compared to similar write-down

Table 5.1 CoCo spreads for different CoCo AT1 Write Down CoCo bonds. The CET1 ratios are Basel III ratios (fully phased-in) of Q3 2015. Their exact publication date is given between brackets. Source: Bloomberg

Issuer	CET1–Q3	ISIN	Cpn	Trigger	CoCo Spread
Credit Suisse	10.2 (Oct 21)	XS0989394589	7 1/2	5.125	481
		XS1076957700	6 1/4	5.125	463
Danske	15.0 (Oct 29)	XS1044578273	5 3/4	7	519
		XS1190987427	5 7/8	7	516
Deutsche bank	11.5 (Oct 29)	DE000DB7XHP3	6	5.125	623
		XS1071551474	6 1/4	5.125	578
		XS1071551391	7 1/8	5.125	568
		US251525AN16	7 1/2	5.125	571
KBC	17.4 (Nov 16)	BE0002463389	5 5/8	5.125	542
Nordea	16.3 (Oct 21)	US65557DAM39	5 1/2	8	466
		US65557DAL55	6 1/8	8	437
		XS1202090947	5 1/4	8	462
Soc. Gen.	10.5 (Nov 5)	XS0867614595	8 1/4	5.125	502
		USF8586CRW49	7 7/8	5.125	566
		XS0867620725	6 3/4	5.125	586
		USF8586CXG25	6	5.125	552
Unicredito	10.53 (Nov 11)	XS1046224884	8	5.125	590
		XS1107890847	6 3/4	5.125	658
Credit Agric.	10.3 (Nov 5)	USF22797RT78	7 7/8	7	539
		XS1055037177	6 1/2	7	567
		XS1055037920	7 1/2	7	553
		USF22797YK86	6 5/8	7	533
UBS	14.3 (Nov 3)	CH0271428333	7	5.125	420
		CH0271428309	5 3/4	5.125	469
		CH0286864027	6 7/8	7	471
		CH0271428317	7 1/8	7	447

bonds issued by the same bank. This is remarkable since in the absence of a valuation model, the market price of the different contingent convertibles are largely in line with similar bonds from the same issuer. In Fig. 5.1, we show the historical implied CET1 volatilities together with the CET1 ratio for a selection of CoCo issuers.

The exceptions are the contingent convertibles from UBS. This issuer has different types of full write-down CoCos outstanding. The two UBS bonds with a high trigger level of 7% have an average implied CET1 volatility of 21.85%. The two CoCo bonds with a low trigger have an average implied CET1 volatility is 28.21%. This difference of more than 6 points in implied volatility for the low-trigger, compared to the high trigger CoCos is in analogy with the observed skew for equity option prices. It is a fact that deep out-of-the-money puts trade at a higher implied volatility than less out-of-the-money options. The same conclusion stands for contingent convertibles.

Table 5.2 Implied CET1 Volatility for different CoCo AT1 Write Down CoCo bonds. The CET1 ratios are Basel III ratios (fully phased-in) of Q3 2015. Source: Bloomberg

Issuer	CET1–Q3	ISIN	Cpn	Trigger	$\bar{\sigma}_{CET1}$
Credit Suisse	10.2 (Oct 21)	XS0989394589	7 1/2	5.125	20.73
		XS1076957700	6 1/4	5.125	20.66
Danske	15.0 (Oct 29)	XS1044578273	5 3/4	7	25.05
		XS1190987427	5 7/8	7	23.82
Deutsche Bank	11.5 (Oct 29)	DE000DB7XHP3	6	5.125	27.09
		XS1071551474	6 1/4	5.125	27.42
		XS1071551391	7 1/8	5.125	25.59
		US251525AN16	7 1/2	5.125	25.66
KBC	17.4 (Nov 16)	BE0002463389	5 5/8	5.125	40.57
Nordea	16.3 (Oct 21)	US65557DAM39	5 1/2	8	23.41
		US65557DAL55	6 1/8	8	20.69
		XS1202090947	5 1/4	8	21.95
Soc. Gen.	10.5 (Nov 5)	XS0867614595	8 1/4	5.125	25.13
		USF8586CRW49	7 7/8	5.125	23.22
		XS0867620725	6 3/4	5.125	24.10
		USF8586CXG25	6	5.125	24.69
Unicredito	10.53 (Nov 11)	XS1046224884	8	5.125	23.86
		XS1107890847	6 3/4	5.125	25.35
Credit Agric.	10.3 (Nov 5)	USF22797RT78	7 7/8	7	13.18
		XS1055037177	6 1/2	7	13.72
		XS1055037920	7 1/2	7	13.61
		USF22797YK86	6 5/8	7	14.04
UBS	14.3 (Nov 3)	CH0271428333	7	5.125	27.02
		CH0271428309	5 3/4	5.125	29.40
		CH0286864027	6 7/8	7	21.15
		CH0271428317	7 1/8	7	22.54

5.2 Adjusted Distance to Trigger

The distance to trigger can now be adjusted by its implied CET1 volatility. We define this adjusted distance to trigger ($D_{\bar{\sigma}_{CET1}}$) as:

$$D_{\bar{\sigma}_{CET1}} = \frac{\log(D)}{\bar{\sigma}_{CET1}}$$

First, we create a simple regression model to determine the CoCo spread from the adjusted distance to trigger using the implied CET1 volatility (see Fig. 5.2). This regression model has an R-squared of 50%. In comparison a similar regression model is constructed but with the unadjusted distance to trigger. The regression model in Fig. 5.3 only gives an R-squared of 2.3%. This result emphasizes that the distance to the trigger is a weak measure to quantify the embedded risk of a contingent

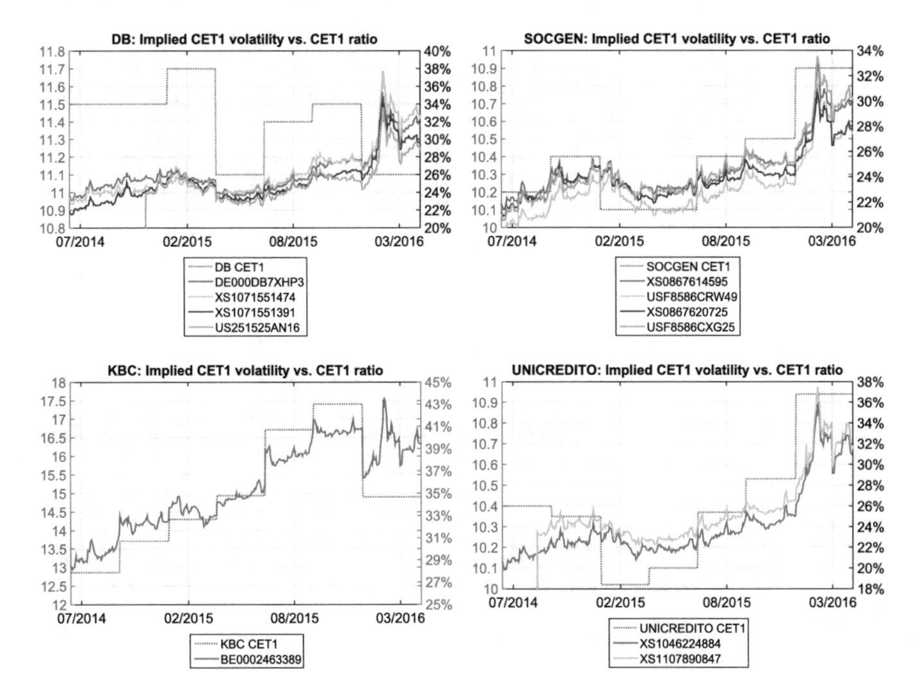

Fig. 5.1 Historical CET1 ratios (left scale) and implied CET1 Volatility (right scale)

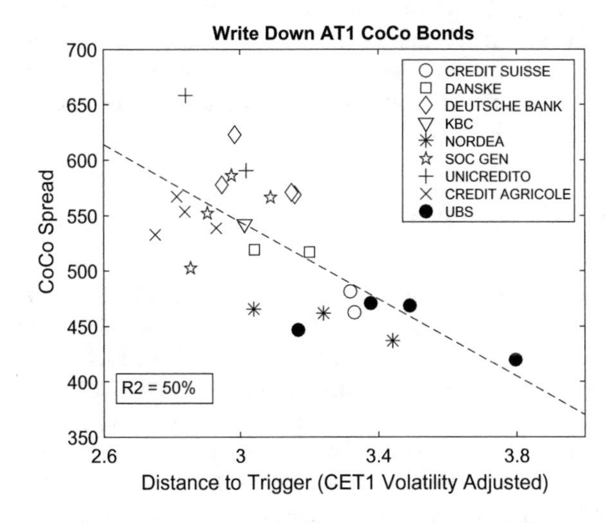

Fig. 5.2 CET1 volatility adjusted distance to trigger versus CoCo spreads

convertible. The CET1 level is a static picture and does not inform a lot about the business risk of a particular financial institution. An increase in non-performing loans, a regulatory fine, unexpected trading losses, etc... could bring the bank into an undesired situation where the CoCo bonds holders see their face value wiped out.

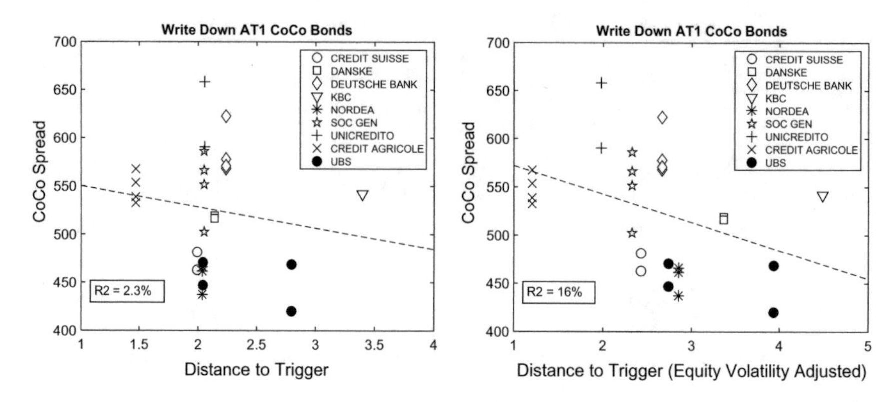

Fig. 5.3 Unadjusted distance to trigger (left) and historical volatility adjusted distance to trigger (right) versus CoCo spreads

Also a regression model is constructed using an equity volatility adjusted distance to trigger. This adjusted distance to trigger using historical equity volatility (D_{σ_S}) is defined by:

$$D_{\sigma_S} = \frac{\log(D)}{\sigma_S}$$

with σ_S the historical volatility of the underlying equity price over a one year horizon. The adjustment based on the equity volatility leads to a slightly increased R-squared of 16% (see Fig. 5.3). However the regression fitting based on the distance to trigger adjusted by the implied CET1 volatility clearly out stands the other regression results.

5.3 Coupon Cancellation Risk

Additional Tier 1 CoCos are more equity-like instruments whereas a Tier 2 CoCo has more bond features. For example an AT1 CoCo is perpetual with its first call date at minimum five years after the issue. Furthermore this call feature cannot include any incentive to redeem earlier. Also, the coupons of an AT1 CoCo can be cancelled without implying a default of the issuer. For Tier 2 CoCos a finite maturity is defined in the prospectus. There is a possibility of a step up at call dates and a coupon cancellation feature is not obligatory for a Tier 2 CoCo.

There is also an outspoken difference observable between the implied volatility levels for the CET1 ratio for T2 and AT1 contingent debt. For example, the Tier 2 and Additional Tier 1 write-down CoCo bonds of Credit Suisse and UBS have almost identical CET1 trigger levels: 5% and 5.125% respectively. Based on the CET1 levels, the market price of the contingent convertibles with a write-down can be used to determine $\overline{\sigma}_{CET1}$. The results of the implied CET1 volatilities are summarised in Table 5.3. The difference in the implied CET1 volatilities equals 2.33% for Credit

Table 5.3 Implied CET1 volatility ($\overline{\sigma}_{\text{CET1}}$) for T2 and T1 bonds issued by Credit Suisse and UBS (May 20th, 2016)

Basel type	CoCo bond (%)	ISIN	Crncy	Trigger	$\overline{\sigma}_{CET1}$	cs
AT1	CS 7.5	XS0989394589	USD	5.125	25.15	554
T2	CS 6.5	XS0957135212	USD	5	22.82	404
AT1	UBS 5.75	CH0271428309	EUR	5.125	30.71	537
T2	UBS 4.75	CH0236733827	EUR	5	28.39	352

Suisse and 2.32% for UBS. This observation can be related to the difference in nature between T2 and AT1 debt. An Additional Tier 1 bond has two extra layers of risk that are not embedded within a Tier 2 structure. First, and most important, is the fact that a Tier 1 bond carries the risk of a coupon cancellation. The issuer has discretion over the coupon payment and such a failure to pay would not trigger a default event. If a coupon of a Tier 2 would be skipped, this would be considered as a failure to pay, causing the issuer to default. A second difference is the fact that AT1 bonds have in theory an unknown maturity. These bonds are perpetual and all have a first call date at least five years after the issue date. The fact that the issuing bank is not obliged to call back the bond on the first occasion introduced an extra layer of risk: extension risk. Both the presence of coupon cancellation and extension risk in the AT1 structures are responsible for the difference in CET1 implied volatility.

Making abstraction from the extension risk, we attribute the difference in implied CET1 volatility to the fact that an investor in the AT1 bond can indeed suffer from a coupon cancellation. The market attributes a higher CET volatility to the AT1 CoCo bond because of this extra layer of risk. This is the main reason why these two categories of CoCo bonds should trade at different implied volatility levels even if their CET1 trigger levels are almost identical. To model this we introduce a new trigger level (CETQ) below which coupons are not paid out.

Figure 5.4 illustrates a hypothetical simulated random walk with an initial CET1 level of 11.4 in May 2016. Suppose, the AT1 CoCo bond distributes a semi-annual coupon in June and December. Coupons will be paid as long as the CET1 level is above the level of the coupon trigger CETQ. The simulation in Fig. 5.4 illustrates how between December 2018 and June 2021, the CET1 fails to remain above the required CETQ level. In this case, the investor would be excluded from 6 consecutive coupon payments. Since the CET1 level remains above the 5.125% trigger, the write-down mechanism is never activated and the investor never suffers from a full write-down.

This allows us to introduce another notion of distance to trigger, or better distance to coupon cancellation. We denote the distance to the CET1 level were the coupons will not be paid out by

$$D_Q = \frac{\text{CET1 ratio}}{\text{CETQ}} \tag{5.2}$$

The pricing formula (Eq. 2.20) of Chap. 2 should now be extended with the coupon cancellation. The coupon component (**Cpn**) will change due to an extra barrier level

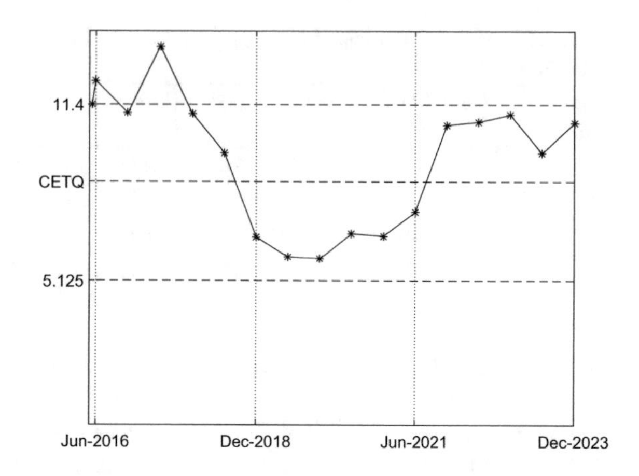

Fig. 5.4 Simulation of random CET1 path for the CREDIT SUISSE 7.5% CoCo Bond

included. The Binary-Down-and-Out options that model the coupons c_i paid out in T_i years time in absence of a trigger and coupon deferral have now two barrier levels, the trigger level and the coupon cancellation level (Rubinstein and Reiner 1991):

$$\mathbf{Cpn} = \sum_{i=1}^{k} \underbrace{c_i \exp(-rT_i)}_{\text{present value } c_i} \times \underbrace{\left[\Phi\left(x_i - \sigma_{\text{CET1}}\sqrt{T_i}\right) - D\Phi\left(y_i - \sigma_{\text{CET1}}\sqrt{T_i}\right)\right]}_{\text{Probability that at time } T_i, D>1 \text{ and } D_Q>1}$$

$$\begin{aligned} &\text{with} \\ x_i &= \frac{\log(D_Q)}{\sigma_{\text{CET1}}\sqrt{T_i}} + \frac{\sigma_{\text{CET1}}\sqrt{T_i}}{2} \\ y_i &= -\frac{\log\left(\frac{D^2}{D_Q}\right)}{\sigma_{\text{CET1}}\sqrt{T_i}} + \frac{\sigma_{\text{CET1}}\sqrt{T_i}}{2} \\ k &= \text{Total number of coupons} \\ D &> D_Q \end{aligned} \tag{5.3}$$

The set of equations above allows us to find the implied level of CETQ for the CREDIT SUISSE 7.5% CoCo Bond using the following input.

CREDIT SUISSE 7.5%	
Name	CREDIT SUISSE 7.5%
ISIN	XS0989394589
Pricing Date	20-May-2016
Coupon	7.5%
Coupon Frequency	Semi-Annual
CET1 Trigger	5.125
T	7.56 yr
CoCo Price (P)	102.34
CET1 Ratio	11.4
σ_{CET}	22.82

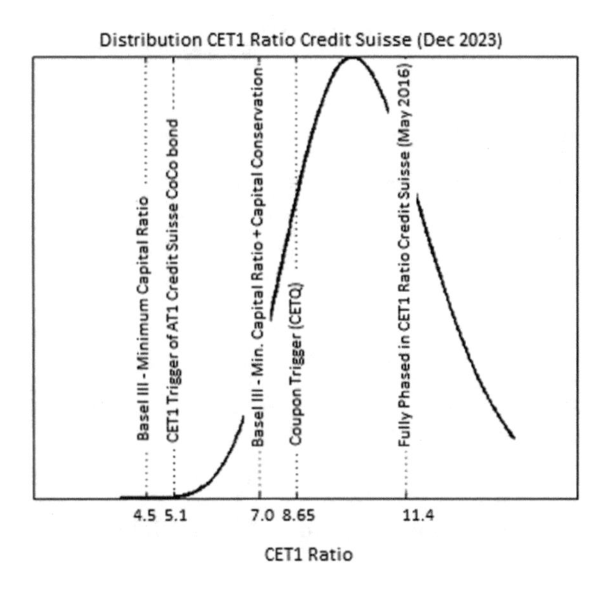

Fig. 5.5 Distribution of the CET1 ratio of Credit Suisse

The theoretical value of this AT1 CoCo bond is priced using the implied CET1 volatility of the CREDIT SUISSE 6.5% (T2) CoCo bond. This volatility is equal to 22.82%. Setting the theoretical value of the CoCo bond as specified in Eq. 5.3 matching with the market price of 102.34, the implied coupon trigger CETQ is equal to 8.65%. This implied level of 8.65% is the CET1 level below which the market considers Credit Suisse will not pay out a coupon on this CoCo bond. We note that this level is higher than the 4.5% minimum CET1 ratio proposed by the Basel Committee plus the required 2.5% capital conservation buffer. Hence the market clearly considers Credit Suisse to cancel a coupon payment on this particular CoCo bond even before the 7% level (4.5% + 2.5%) is reached. In Fig. 5.5 the distribution of the CET1 ratio of Credit Suisse was graphed using Eq. (2.16) with a CET1 volatility of 22.82%. The time horizon was extended up till the first call date of the bond in December 2023.

The presence of a coupon cancellation has a negative impact on the value of the contingent convertible. To illustrate this impact the cancellation risk for the Credit Suisse CoCo bond is defined as the percentage impact on the theoretical value of the coupon bond due to the presence of the coupon trigger CETQ:

$$\text{Cancellation Risk} = \frac{\text{Cpn}_{CETQ=5.125} - \text{Cpn}_{CETQ=8.65}}{P_{\text{Theo}}}$$

The different values for Cpn are calculated using Eq. 5.3. The results for May 20, 2016 are summarised in Fig. 5.6. From this figure, we learn that the coupon cancellation risk of the AT1 Credit Suisse CoCo represents around 10.4% of the CoCo bond's market price.

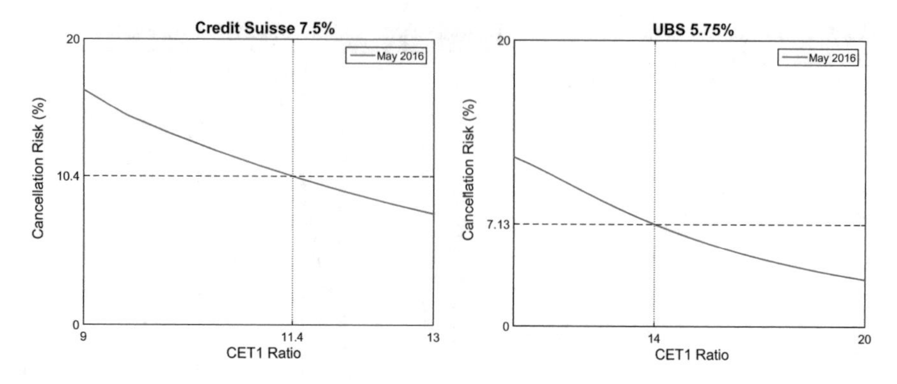

Fig. 5.6 Percentage impact of the presence of a possible coupon cancellation on theoretical value of the Credit Suisse (left) and UBS (right) AT1 CoCo Bond

Similar results can be found for the UBS CoCo given the following input.

UBS 5.75%	
Name	UBS 5.75%
ISIN	CH0271428309
Pricing Date	20-May-2016
Coupon	5.75%
Coupon Frequency	Annual
CET1 Trigger	5.125
T	5.76 yr
CoCo Price (P)	101.18
CET1 Ratio	14
σ_{CET}	28.39

We find an implied coupon trigger level at 9.83% corresponding with a coupon cancellation risk represented around 7.13% of the UBS CoCo market price as shown in the right graph of Fig. 5.6.

5.4 Conclusion

Contingent Convertible bonds can be seen as derivative instruments contingent on the CET1 level. The unadjusted distance to the trigger, i.e. the distance between the current CET1 ratio and its trigger level, is a weak measure to quantify the embedded risk of a contingent convertible. The CET1 level is a static picture and does not inform a lot about the business risk of a particular financial institution. The notion of implied CET1 volatility is introduced and used to define a risk-adjusted distance to trigger. The CET1 volatility adjusted distance to trigger has much more explanatory power in describing CoCo spreads than pure distance to trigger.

As a conclusion different contingent convertibles issued by the same bank and sharing a similar CET1 ratio, have almost identical implied CET1 volatility levels. The same results confirm the difference in market risk between Tier 2 and Additional Tier 1 CoCo bonds. The ability to obtain an implied level for the CET1 volatility offers also an other interesting result. The implied coupon cancellation level can be estimated.

Chapter 6
Outlier Detection of CoCos

Data mining is an important new topic within the financial world with multiple applications in risk management, trading, marketing etc. For example banks apply data mining in various areas from credit scoring to the pricing of loans. In this chapter the focus is on the detection of observations different from the majority, called outliers. This can be of interest for market analysts, risk managers, regulators and traders. The exceptions might be caused by exceptional circumstances and can require extra hedging or can be seen as trading opportunities. They could as well give regulators an early warning and signal for potential trouble ahead.

In this section the risk of contingent convertible bonds is measured and compared with other asset classes. We explain and apply the new risk measure, called the Value-at-Risk Equivalent Volatility (VEV), to different CoCos. The concept was introduced by the European authorities in the new regulation for Packaged Retail and Insurance-based Investment Products (PRIIPs) and has to be implemented for all structured products since 2018. This risk-measure is an extension of the classical volatility measure by taking into account skewness and kurtosis.

On the financial markets, extreme CoCo price movements occur in general when the underlying equity prices move. This relation is clear by construction of the CoCo asset class. However to detect outliers in the CoCo market one should take into account multiple variables such as the CoCo market returns and the underlying equity returns. A robust measure for the autocorrelation is defined to detect outlying behaviour in a multi-dimensional setting. Based on this distance, CoCos can be detected that are outlying compared to previous time periods but taking into account extreme moves of the market situation as well.

In the first part we measure the risk of CoCos in terms of VEV. With this measure we will also be able to recognize the hybrid character of the CoCos. In the second part, a multivariate setting is applied where the covariance between the stock market and the CoCo market will be taken into account. The robust method of Minimum Covariance Determinant (MCD) is explained in order to detect outlying CoCos. More information can be found in De Spiegeleer et al. (2017).

J. De Spiegeleer et al., *The Risk Management of Contingent Convertible (CoCo) Bonds*, SpringerBriefs in Finance, https://doi.org/10.1007/978-3-030-01824-5_6

6.1 Value-at-Risk Equivalent Volatility (VEV)

Financial analysts are often referring to volatility as a risk metric expressing the uncertainty in the returns in their products and portfolios. A CoCo typically sits between debt an equity in terms of volatility. This confirms the hybrid nature of the CoCo bond. Also preferreds are exhibiting similar volatility. A volatility cone displays these historical volatility values for multiple window sizes. The cone is constructed out of 90% (and 10%) upper (resp. lower) bounds for the volatility. These boundaries come closer to each other for longer windows due to the diversification of the returns. The risk of the CoCo bond is at an intermediate level in between the cone of the equity and bond (Fig. 6.1).

Regulators want clarity and transparency for the financial instruments offered to investors. Pre-contractually, a retail investor receives for a Packaged Retail and Insurance-based Investment Product (PRIIP), a simple document, called Key Information Document (KID), with clear facts and figures on the risks of a particular financial instrument. The new technical standards classify PRIIPs using a new indicator called the 'Summary Risk Indicator' (SRI). This integer number takes values in a range from one to seven. Market and credit risk are taken as the major factors of risk that need to be reflected in this indicator, alongside liquidity risk (European Commission 2017). The determination of the market risk relies on the concept of Value-at-Risk Equivalent Volatility (VEV). VEV is calculated based on the Value-at-Risk (VaR) levels of a product and translates this value back to the concepts of volatility.

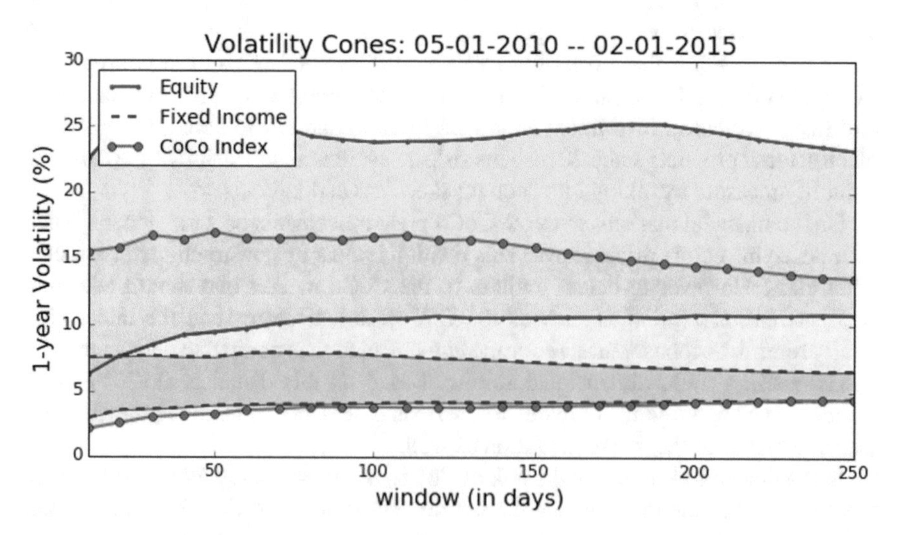

Fig. 6.1 The volatility cone displays the 10 until 90% annualized volatility for multiple window sizes (in days). The cone displays the range of risk for the different asset classes: iShares Core S&P 500 equity ETF, iShares iBoxx Investment Grade fixed income ETF and Credit Suisse CoCo Index

The Value-at-Risk Equivalent Volatility (VEV) is a new risk measure to evaluate different assets. The VEV denotes the volatility that corresponds with a Value-at-Risk (VaR) loss event. For the PRIIPs, the market risk is measured by the annualised volatility corresponding to the VaR measured at the 97.5% confidence level over the recommended holding period unless stated otherwise. The VEV formula is a closed form formula to evaluate the risk.

Under the Black–Scholes model, we know that stock prices are lognormally distributed with drift μ and σ volatility. It follows that the logreturns over a time period T becomes distributed as:

$$r_T \sim N\left[\mu - \frac{\sigma^2}{2}T; \sigma^2 T\right]. \tag{6.1}$$

For a zero drift, we find the following VaR formula:

$$P(r_T < VaR_{1-\alpha}) = \alpha \tag{6.2}$$

$$P\left(Z < \frac{VaR_{1-\alpha} + \frac{\sigma^2}{2}T}{\sigma\sqrt{T}}\right) = \alpha \tag{6.3}$$

$$\frac{VaR_{1-\alpha} + \frac{\sigma^2}{2}T}{\sigma\sqrt{T}} = -z_\alpha \tag{6.4}$$

$$VaR_{1-\alpha} = -\frac{\sigma^2 T}{2} - z_\alpha \sigma\sqrt{T} \tag{6.5}$$

with volatility σ, time period T and $z_\alpha = N^{-1}(1-\alpha)$, the $100(1-\alpha)\%$-percentile of a standard normal distribution. For $\alpha = 2.5\%$, the z-score equals $z_\alpha = 1.96$ and the value-at-risk becomes:

$$VaR_{97.5\%} = -\frac{\sigma^2 T}{2} - 1.96\sigma\sqrt{T} \tag{6.6}$$

A lot of literature has been created on the disadvantages of the normal distribution as model for financial returns on different markets and periods (see for example Mandelbrot 1963 and Brinner 1974). In this perspective the traditional volatility (e.g. used above in the volatility cone) is not a perfect risk measure. Volatility is a symmetric measure by means of treating returns above and below the expected return equally. However, in practice most logreturns turn out to be asymmetrically distributed and have fatter tails, which is referred to in statistics by resp. skewness and kurtosis. These measures can be derived from the central moments, denoted by $M_k = E[(R - E(R))^{k+1}]$ for $k = 1, 2, 3$. The variance, skewness and excess kurtosis of the logreturns become:

$$V = \sigma^2 = M_1 \tag{6.7}$$

$$S = \frac{M_2}{\sigma^3} \tag{6.8}$$

$$K = \frac{M_3}{\sigma^4} - 3 \tag{6.9}$$

Over N trading periods, we have a sample mean of (daily) returns R_i, defined by:

$$\bar{R} = \frac{1}{N}\sum_{i=1}^{N} R_i \tag{6.10}$$

The period size N is referred to as the recommended holding period. The variance, skewness and kurtosis of this sample mean become:

$$V_N = \frac{V}{N} \tag{6.11}$$

$$S_N = \frac{S}{\sqrt{N}} \tag{6.12}$$

$$K_N = \frac{K}{N} \tag{6.13}$$

To extend the VaR for non-normal distributed variables, the regulation applies the Cornish-Fisher expansion (Cornish and Fisher 1938). This expansion estimates quantiles of non-normal distribution based on the first four moments by:

$$VaR_{1-\alpha} = -\frac{\sigma^2 N}{2} + \sigma\sqrt{N}\left(-z_\alpha + (z_\alpha^2 - 1)\frac{S_N}{6} + (-z_\alpha^3 + 3z_\alpha)\frac{K_N}{24} - (-2z_\alpha^3 + 5z_\alpha)\frac{S_N^2}{36}\right) \tag{6.14}$$

$$VaR_{1-\alpha} = -\frac{\sigma^2 N}{2} + \sigma\sqrt{N}\left(-z_\alpha + (z_\alpha^2 - 1)\frac{S}{6\sqrt{N}} + (-z_\alpha^3 + 3z_\alpha)\frac{K}{24N} - (-2z_\alpha^3 + 5z_\alpha)\frac{S^2}{36N}\right) \tag{6.15}$$

For $\alpha = 2.5\%$, VaR becomes:

$$VaR_{1-\alpha} = -\frac{\sigma^2 N}{2} + \sigma\sqrt{N}\left(-1.96 + 0.4736\frac{S}{\sqrt{N}} - 0.0687\frac{K}{N} + 0.1461\frac{S^2}{N}\right) \tag{6.16}$$

The 1-year VaR Equivalent Volatility (VEV) is the volatility parameter in Formula (6.6) where the VaR is given by Formula (6.15). This results in:

$$VEV = (\sqrt{z_\alpha^2 - 2VaR_{1-\alpha}} - z_\alpha)/\sqrt{T}, \tag{6.17}$$

with T the length of the recommended holding period in years.

Remark that the VEV for a normal distribution (S = 0 , K = 0) becomes:

$$VEV = \sigma\sqrt{\frac{N}{T}}$$

If the recommended holding period is equal to N trading days, the length T equals to $N/250$ years. As such, under the normal distribution, we can rescale the daily volatility to a 1-year VEV by multiplying with factor $\sqrt{250}$.

6.1.1 Common Pitfalls

The closed form formula (Eq. 6.17) can be evaluated for each financial instrument based on the first four moments of its returns. However the measure remains a backward looking approach since these moments will be derived from the historical time series. Attention should also be given to the defining of the recommended holding period. A longer recommended holding period reduces the skew and kurtosis parameter. As such the VEV becomes closer to the volatility. Also the application of the Cornish-Fisher expansion needs extra attention for the user. The skewness and kurtosis values in the expansion are parameters and do not coincide with the actual kurtosis and skewness of the instrument. Furthermore the formula is only applicable in a certain range of parameter values which is referred to as the domain of validity.

Recommended Holding Period

There is a major impact on the VEV value by changing the recommended holding period of the PRIIP. The European Supervisory Authorities (incl. ESMA, EIOPA and EBA) point out that a brief description should be given of the reasons for the selection of the recommended holding period and, where present, the required minimum holding period. In addition each KID document has to indicate that the risk of the product if not held to maturity or for the recommended holding period may be significantly higher than the one represented in the report (European Commission 2017).

In a case study we show the impact on the VEV risk measure due to a change in recommended holding period. The study investigates 103 CoCos consisting of 82 AT1 CoCos and 21 Tier 2 CoCo. In Table 6.1 we display the number of CoCos issued per bank.

Assume the recommended holding period of the CoCos is one year, the number of trading days in this holding period is $N = 250$. As such the VaR in the calculation becomes a 1-year VaR based on an average 1-year skew and kurtosis. The rescaling factor (T) equals one because no rescaling is necessary in order to express the VEV over a term of one year. On the other hand, if the recommended holding period of the CoCo is equal to one trading day ($N = 1$), the skew and kurtosis are not rescaled in Eqs. 6.12 and 6.13. As a result the value for the VaR is a 1-day VaR based on 1-day skew and kurtosis. Here we should take the rescaling factor $T = 1/250$. Notice that

Table 6.1 Scope of CoCos: Number of CoCos per issuer

Issuer	AT1	T2	Total
UBS	4	5	9
BACR	6	2	8
CS	3	4	7
LLOYDS	6	0	6
ACAFP	5	1	6
SOCGEN	5	0	5
HSBC	5	0	5
RABOBK	3	1	4
DB	4	0	4

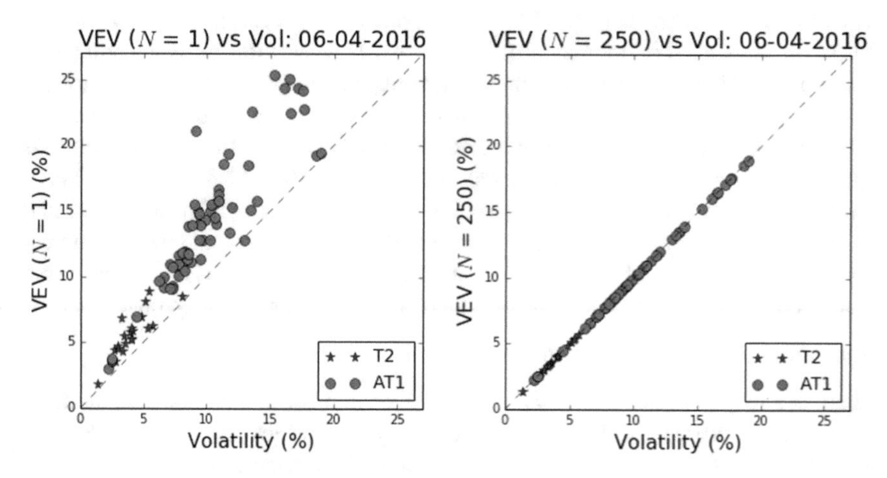

Fig. 6.2 The impact of the recommended holding period on the VEV. A 1-year VEV is displayed with respect to the 1-year volatility (in %) on April 6, 2016 for $N = 1$ (left) and $N = 250$ (right)

there is a major difference in the values of the VEV over a one year term (see Fig. 6.2). The largest difference is observed for the Credit Suisse CoCo (CS 6 09/29/49) which has a VEV of 21.13% and a volatility of 9.11% on April 6, 2016. The impact of non-normal distribution is best observed for a recommended holding period of one day (or one trading period). This maximizes the difference between both measures. Therefore we apply $N = 1$ in the next sections.

Application of Cornish-Fisher Expansion

The Cornish-Fisher expansion describes a transformation of a normal distributed variable to a non-normal distributed variable by using the first four moments of the non-normal variable. Suppose Z is a standard normal distributed variable. The transformation is then given by:

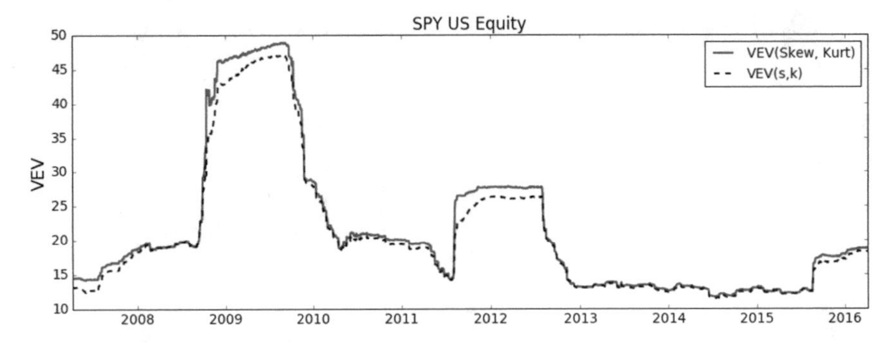

Fig. 6.3 Difference in the VEV for the S&P 500 ETF price based on the skewness and kurtosis parameters and the actual skewness and excess kurtosis

$$X = Z + (Z^2 - 1)\frac{s}{6} + (Z^3 - 3Z)\frac{k}{24} - (2Z^3 - 5Z)\frac{s^2}{36} \tag{6.18}$$

where X is a non-normal distributed variable. The parameters s and k can be derived from the actual skewness and excess kurtosis of X. The following relation holds between the central moments of X and the parameters s and k (Maillard 2012):

$$M_1 = 0 \tag{6.19}$$

$$M_2 = 1 + \frac{1}{96}k^2 + \frac{25}{1296}s^4 - \frac{1}{36}ks^2 \tag{6.20}$$

$$M_3 = s - \frac{76}{216}s^3 + \frac{85}{1296}s^5 + \frac{1}{4}ks - \frac{13}{144}ks^3 + \frac{1}{32}k^2 s \tag{6.21}$$

$$M_4 = 3 + k + \frac{7}{16}k^2 + \frac{3}{32}k^3 + \frac{31}{3072}k^4 - \frac{7}{216}s^4 - \frac{25}{486}s^6 + \frac{21665}{559872}s^8 \tag{6.22}$$

$$- \frac{7}{12}ks^2 + \frac{113}{452}ks^4 - \frac{5155)}{46656}ks^6 - \frac{7}{24}k^2s^2 + \frac{2455}{20736}k^2s^4 - \frac{65}{1152}k^3s^2 \tag{6.23}$$

Notice that the parameter values do not coincide with the effective skewness and kurtosis except for very low values. However the skewness and excess kurtosis are in general not close to zero for financial timeseries since the returns are in general not normal distributed. The impact of the difference is displayed in Fig. 6.3. This is the first common pitfall of the application of the Cornish-Fisher expansion.

The second pitfall is the domain of validity. The formula only holds for a specific range of skewness and kurtosis values. The domain is derived to conserve the order of the quantiles of the distribution. In Maillard (2012), the domain of validity is described in terms of the parameters k and s by:

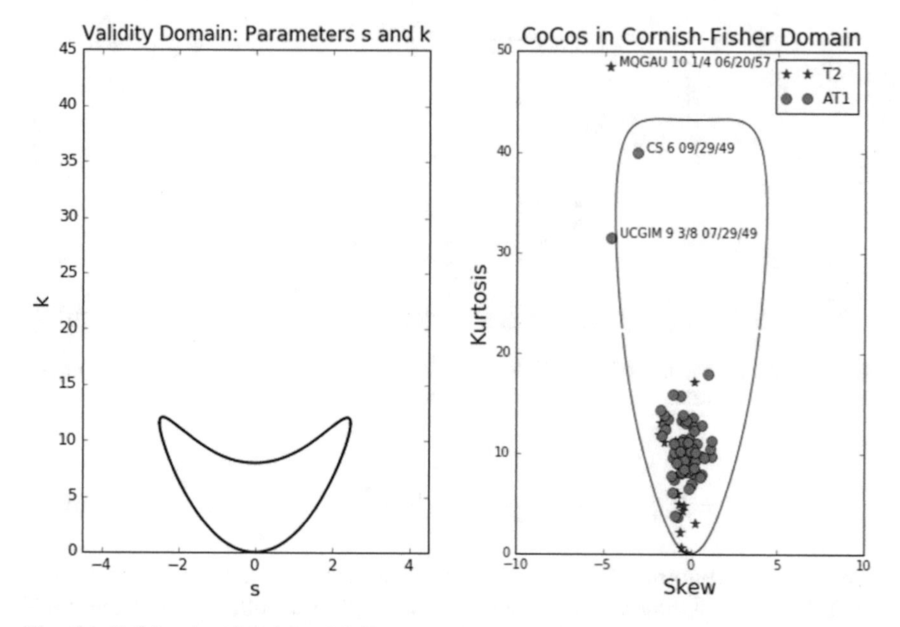

Fig. 6.4 Validity domain of Cornish-Fisher expansion in terms of the parameters (left) and the actual skewness and excess kurtosis (right). A few CoCos are situated outside the validity domain on April 6, 2016

$$-6(\sqrt{2}-1) \le s \le 6(\sqrt{2}-1) \tag{6.24}$$

$$4+\frac{11}{9}s^2-\frac{4}{6}\sqrt{\frac{1}{36}s^4-6s^2+36} \le k \le 4+\frac{11}{9}s^2+\frac{4}{6}\sqrt{\frac{1}{36}s^4-6s^2+36} \tag{6.25}$$

This domain is displayed in Fig. 6.4. The boundaries can be translated in terms of the actual skewness and kurtosis. The CoCos of Table 6.1 were added to check the applicability of the VEV formula. We see that in general most CoCos are situated in the validity domain.

6.1.2 *Case Study: Risk of Different Asset Classes*

In this section a case study is provided to observe differences in risk among the different asset classes. We compare the risk parameters volatility and VEV of multiple Exchange Traded Funds (ETFs) with the Credit Suisse CoCo index. The dataset consist of 20 equity ETFs, 14 mixed ETFs and 21 fixed income ETFs. We observe the values for the VEV on April 6, 2016 and take an observation window of 250 business days. In Fig. 6.5, a Gaussian kernel density estimator (KDE) gives an overview of how the realized volatilities are distributed. Even with a regulatory measure that takes

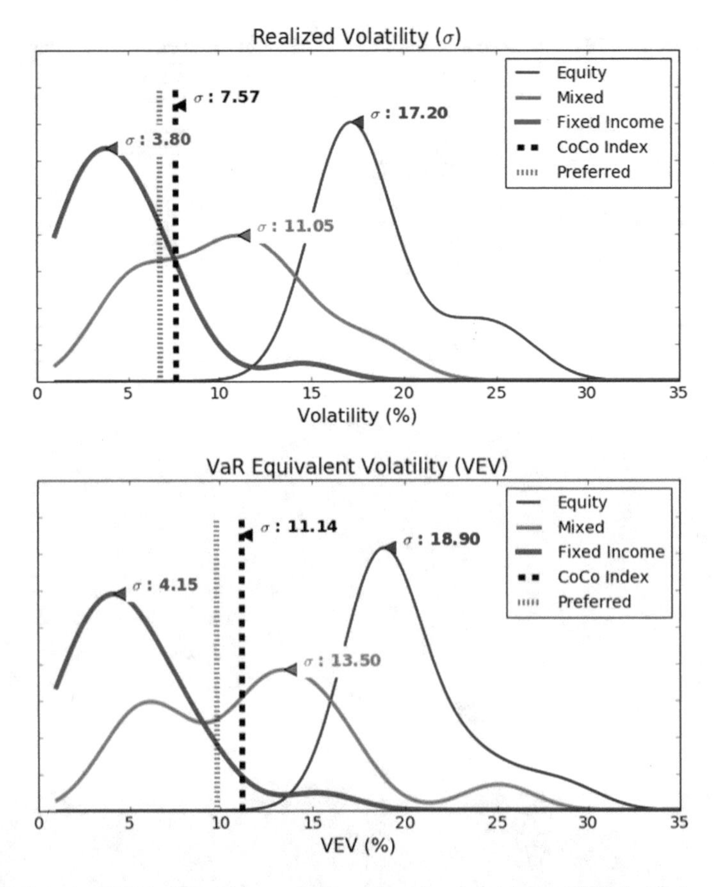

Fig. 6.5 Gaussian KDE of the 1-year volatility (above) and the 1-year VEV (below) per type of asset class compared with the Credit Suisse CoCo Index on April 6, 2016

the fat-tail risk of CoCo bonds into account, the risk profile of CoCos remains distant from equity risk. The risk of the preferred 'Ishares US Preferred Stock' seems to be slightly lower than the CoCo index risk. The question raises if in times of stress the VEV of CoCos can enter in the range of the equity class type.

In European Commission (2017) each asset (PRIIP) is assigned to a specific market risk measure (MRM) class according to its VEV value. From the distributions of previous exercise, we can observe which MRM class is best fitted for each asset type (see Table 6.2). The MRM class should be increased by 1 additional level if the PRIIP is only having monthly price data. The CoCo index does belong to the market risk category 3 together with the Preferred ETF and the Mixed ETFs. This does not mean their is no difference in their risk profiles. For Deutsche Bank and the Banco Popular CoCo, the VEV moves up to risk category 6 (see Fig. 6.6).

Table 6.2 Each PRIIP is classified in a Market Risk Category based on its VEV value

MRI	VEV(%)	Asset type
1	<0.5	–
2	0.5–5.0	Fixed Income ETF
3	5.0–12.0	Preferred, CoCo (index) AND Mixed ETF
4	12.0–20.0	Equity ETF
5	20.0–30.0	–
6	30.0–80.0	–
7	>80.0	–

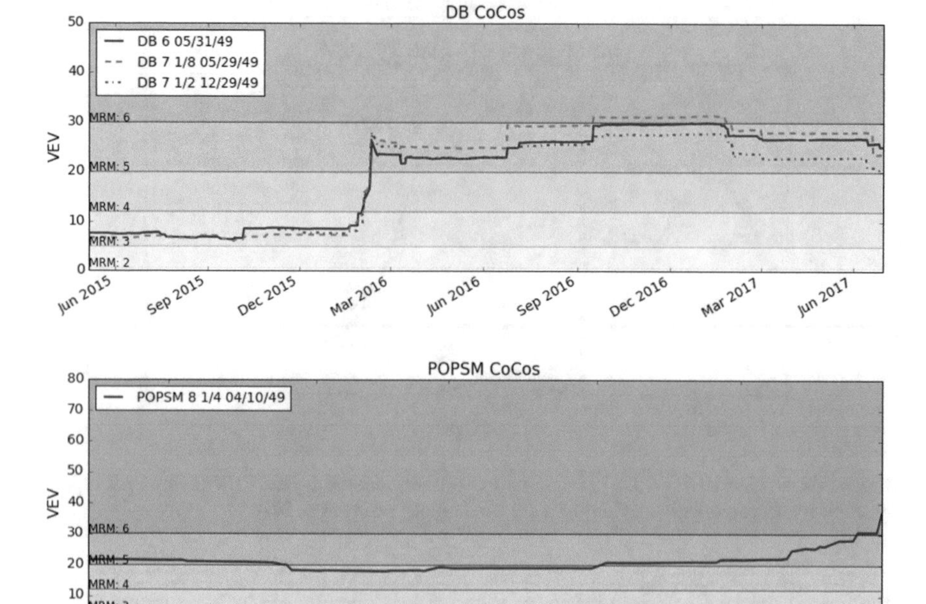

Fig. 6.6 Moves of the VEV of the Deutsche Bank and Banco Popular CoCos over multiple market risk categories

6.2 Are CoCos Moving Out of Sync?

Market analysts are often thinking about risk in terms of sigma-events. These events can be translated in a frequency of occurrence as shown in Table 6.3. But in 2008 markets were observing events that were 25-standard deviation events and occurring several days in a row. Also in terms of the daily returns of the first quarter of 2016,

Table 6.3 Risk in terms of sigma-events

Sigma	Frequency	Explanation
1σ	1 in 3	Twice a week
2σ	1 in 22	Monthly
3σ	1 in 370	Every year and a half
4σ	1 in 15,787	Twice a lifetime
5σ	1 in 1,744,278	Once a history (5000 years)

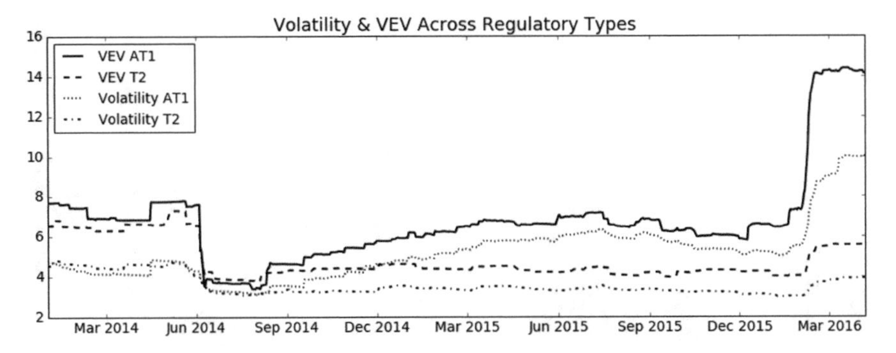

Fig. 6.7 Average volatility and VEV for AT1 CoCos and T2 CoCos

the CoCo asset class observed a real-extreme situation compared to the returns of 2015.

The sigma-events are typically related to z-scores (standardized values) which are often used for univariate outlier detection for continuous variables. In Fig. 6.7, we show the average CoCo volatility and VEV over time for the AT1 CoCos and T2 CoCos. A clear increase in both risk measures is visualised during the first quarter of 2016. The asymmetric tail risk is causing a higher increase in the VEV risk measure compared to the volatility.

The real issue is however if the CoCo bonds are behaving outside the risk defined in the contract such as their sensitivity with the underlying share price return. Did something clearly broke down at the start of 2016? Or were CoCos following the price performance of the bank shares? To see whether Q1 2016 was an outlier, we should not look only at the CoCo bond returns but take into account the share price returns as well. In Fig. 6.8, we show a scatterplot of the daily returns of the Credit Suisse CoCo index versus the Stoxx Banking Index for the different years. Instead of using a volatility measure in one dimension (σ) we will use a covariance matrix Σ of the equity returns and CoCo market returns. In the next sections, we explain our approach in a higher dimensional space and apply it to detect outliers in the CoCo market.

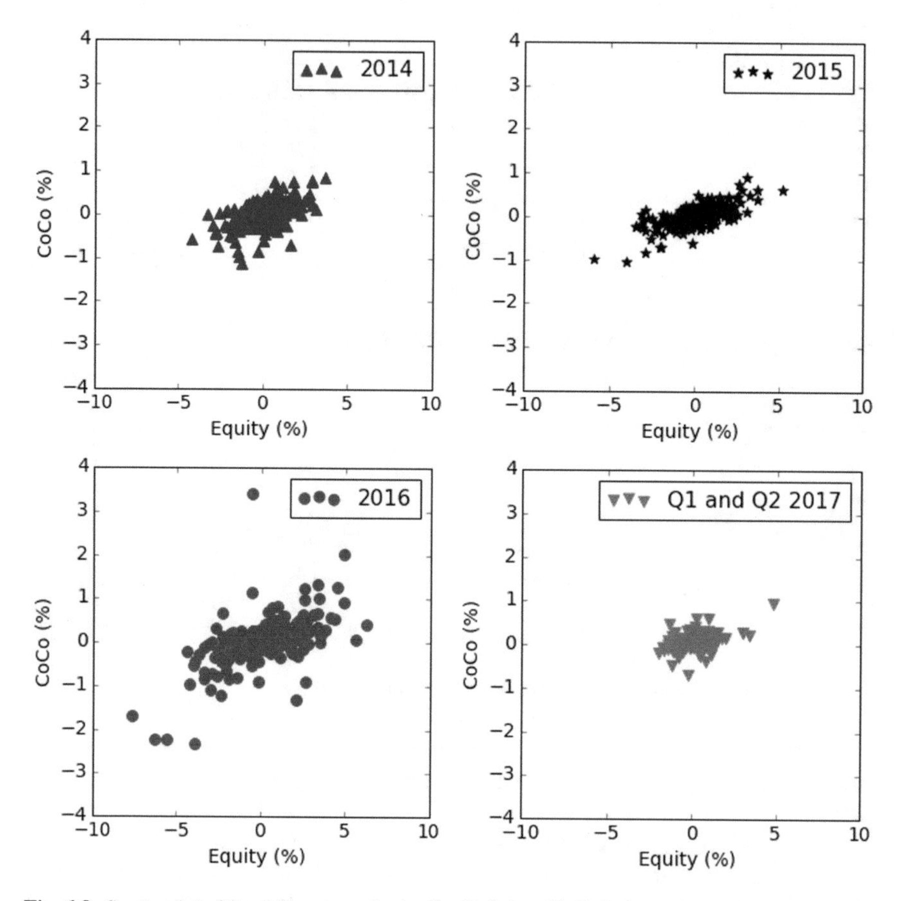

Fig. 6.8 Scatterplot of the daily returns in the Credit Suisse CoCo index versus daily returns in the Stoxx Banking Index

6.2.1 Minimum Covariance Determinant (MCD)

Outliers are in a multivariate setting no longer defined as a z-score but as a Mahalanobis Distance (MD). This distance measures a point x versus a data cloud X as defined by:

$$MD_X(x) = \sqrt{(x - \mu_x)\Sigma^{-1}(x - \mu_x)^T} \tag{6.26}$$

where Σ denotes the covariance matrix of X. Intuitively, the $x - \mu_X$ shows how far a point x stands away from the center of the cloud. In the meantime Σ explains the spread on the dataset X. The distance is hence based on the correlation between the variables. It measures the connectedness of two sets with multiple variables. The distance reduces to the Euclidean distance if the covariance matrix is the identity matrix, and the normalised Euclidean distance if the covariance matrix is diagonal.

The MD measures how many sigma-events a data point is away from the center of a multivariate distribution (Hoyle et al. 2016).

The Mahalanobis distance can be used to find outliers in multivariate data. The squared Mahalanobis distance is chi-squared distributed with p degrees of freedom under the assumption that the p-dimensional dataset is multivariate normal distributed. For a sample of size n, we denote each observation by $x_i \in R^p$ with $i = 1, ..., n$. The estimated Mahalanobis distance is denoted with $MD_X(x_i)$. Afterwards the squared MD is compared with the quantiles of the chi-squared distributed with p degrees of freedom. For example, if the squared MD is larger than the 99% quantile, the observation can be classified as a potential outlier.

Notice that this distance measure is very sensitive to outliers itself. Single extreme observations, or groups of observations, departing from the main data structure can have a severe influence to this distance measure. Both the location and covariance are usually estimated in a non-robust manner. In order to provide reliable measures for the recognition of outliers, one should apply a more robust measure for location and covariance. In practice classical fitting methods used to detect outliers are often so strongly affected by the outliers that the resulting fitted model does not allow to detect deviating observations. This phenomenon is called the masking effect (Rousseeuw et al. 2006).

Different solutions exist to make the distance measure less influenced by outliers or more robust. One approach is to apply the Minimum Covariance Determinant (MCD) method. MCD is a commonly-used robust estimate of dispersion which can be used to construct robust MDs. The MCD estimator is computationally fast algorithm introduced in Rousseeuw and van Driessen (1999).

Using robust estimators of location and scatter in formula (6.26) leads to so-called robust distances. In Rousseeuw and van Zomeren (1990), the robust MD is used to derive a measure of outlyingness. First part in the derivation of the robust Mahalanobis distance is the concentration step. The dataset is divided in different non-overlapping subsamples. For each subsample it computes the mean and the covariance matrix in each feature dimension of the subsample (Hubert and Debruyne 2009). The MD is computed for every multidimensional data vector x_i. Afterwards, the data are ordered ascendantly by this distance in each subsample. Next, subsamples with the smallest MD are selected from the original samples. This procedure is iterated until the determinant of the covariance matrix converges (see Hoyle et al. 2016). Hence the robust measure first selects a subset of the original data whose classical covariance has the lowest determinant. The determinant of a covariance matrix indicates how much space the data-cloud takes. Second part is a correction step to compensate the fact that the estimates were learned from only a portion of the initial data (Pison et al. 2002). This robust Mahalanobis distance also assumes a multivariate normal distributed dataset and does not account for the sample size of the data.

Hardin and Rocke (2005) showed that the cut-off value derived from the chi-square distribution (i.e. $\chi^2_{p,(1-\alpha)}$) is based on the asymptotic distribution of the robust distances. This often indicates too many observations as outlying which means that test results show more false-positive detections than expected for robust MD. In Hardin and Rocke (2005) the corrected distribution of the robust distances is approximated

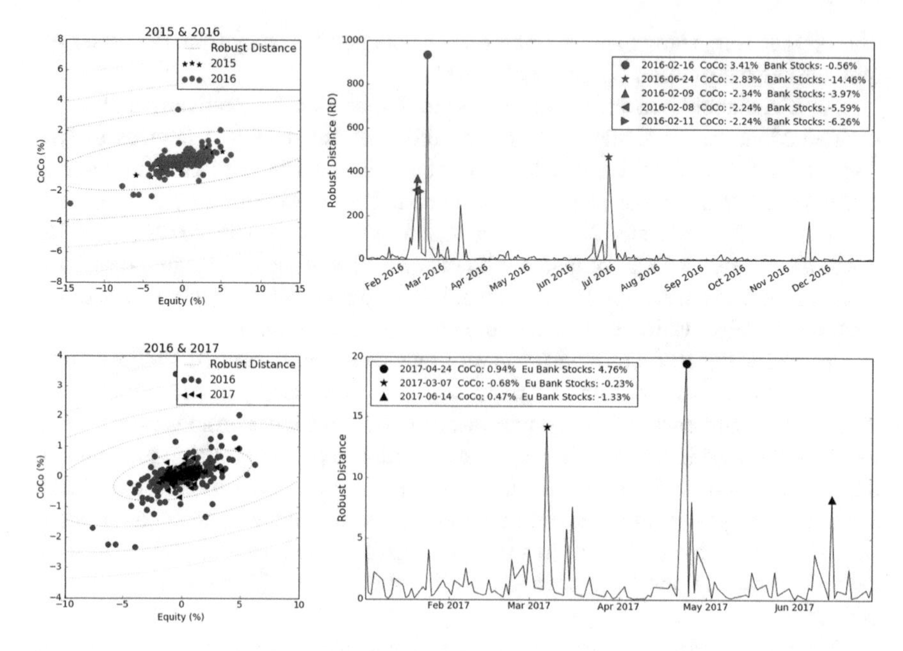

Fig. 6.9 Left: Scatterplot of the daily returns in the CoCo Credit Suisse Index versus daily returns in the Stoxx Banking Index (left) and the Robust Mahalanobis Distance (right)

by the following F-distribution:

$$\frac{np}{1-p} MD_X^2(x) \sim F_{p,n-p} \tag{6.27}$$

Cerioli (2010) also rejects the chi-square quantiles for the detection of outliers. The author developed a new calibration methodology, called Iterated Reweighted MCD (IRMCD), which provides outlier detection tests with the correct Type I error behaviour for the robust MD. In Green (2014) an extension is developed that combines the method of Hardin and Rocke with the IRMCD method of Cerioli.

6.2.2 Measuring the Outliers

An application of this method is the study of irregular behaviour in the relationship between equity returns versus CoCo bond returns. The detection of irregular behaviour will guide us to possible dislocations and potential stability risks.

Outliers Compared to Previous Year

We compare the daily returns in year T with the previous year $T-1$. In Fig. 6.9, we display the robust Mahalanobis distance over the year 2016 (resp. 2017) where we train the data on the returns of stock price and the CoCo price index in the year before: 2015 (resp. 2016). The data points with an extreme distance compared to the overall group are marked and mentioned in the legend.

In February 2016 a general fear over the Europe's banking industry was observed and concerns were raised about Deutsche Bank's ability to pay off the high coupon values of CoCos. During certain days the CoCos move extremely compared to the historical CoCo price returns and their underlying equity returns. In June 2016 the outlier detection method highlights the Brexit election as an outlying phenomenon.

Also the outliers of 2017 can be related to market circumstances. The first outlier in March 2017 was caused by UniCredit due to uncertainty in its next AT1 coupon payment. On April 24, 2017 the outcome of the French election lighted up the EU Bank Stoxx indicating a second outlier. The latest outlier in June 2017 is probably related with the UBS shares drop down due to concerns over margins in its wealth management division. This impacted the Stoxx Banking index whereas the overall CoCo prices remained stable. The general CoCo market proves resilient while losses were imposed on Banco Popular bondholders in June 2017.

Outlier Detection per Issuer

In a next step we detect outlying behaviour of CoCos from specific issuers. The model is fit to the CoCo price return and the underlying equity return during a 90-day history window. In Fig. 6.10, the averaged daily robust MD is shown for different issuers. During certain days these observations move extremely compared to the previous 90-day time period.

Deutsche Bank had to reassure its coupon payments of its outstanding CoCos during the first quarter of 2016. Cancellation of the high coupons in a CoCo would be a significant loss for the CoCo investor. This is also observed in the robust distance of Deutsche Bank that moves for a longer time period out of the boundary derived from the 99% quantile of the F-distribution. In the beginning of 2016, also other CoCo distances move out of this boundary. In February 2016 the CoCo market did experience large losses. During certain days the CoCos moved extremely compared to the historical CoCo price returns and their underlying equity returns.

On June 6, 2017 the European Central Bank considered the bank Banco Popular as "failing or likely to fail" (European Central Bank 2017). This classification is used by supervisors to indicate institutions that become non viable. The Single Resolution Board stepped in forcing the sale to Banco Santander. As part of the deal Banco Popular's junior bonds were wiped out including its CoCo bonds. That marks the first write-off of CoCos industry-wide since regulators developed the bonds in the wake of the financial crisis. The sale spared Spain's taxpayers the cost of another bailout. Banco Popular's CET1 fully loaded ratio, stood at 7.33 percent in March, one of the weakest among European lenders. The remaining market for AT1 bonds remained stable after the take-over.

This corresponds with the outlying behaviour of the Banco Popular (POPSM) CoCo. The robust distance detects periods of stress for the bank starting in mid 2015. Also during the first quarter of 2016, Banco Popular remains outside its boundaries for a longer time period compared to other banks. On April 3, 2017 the robust distance of

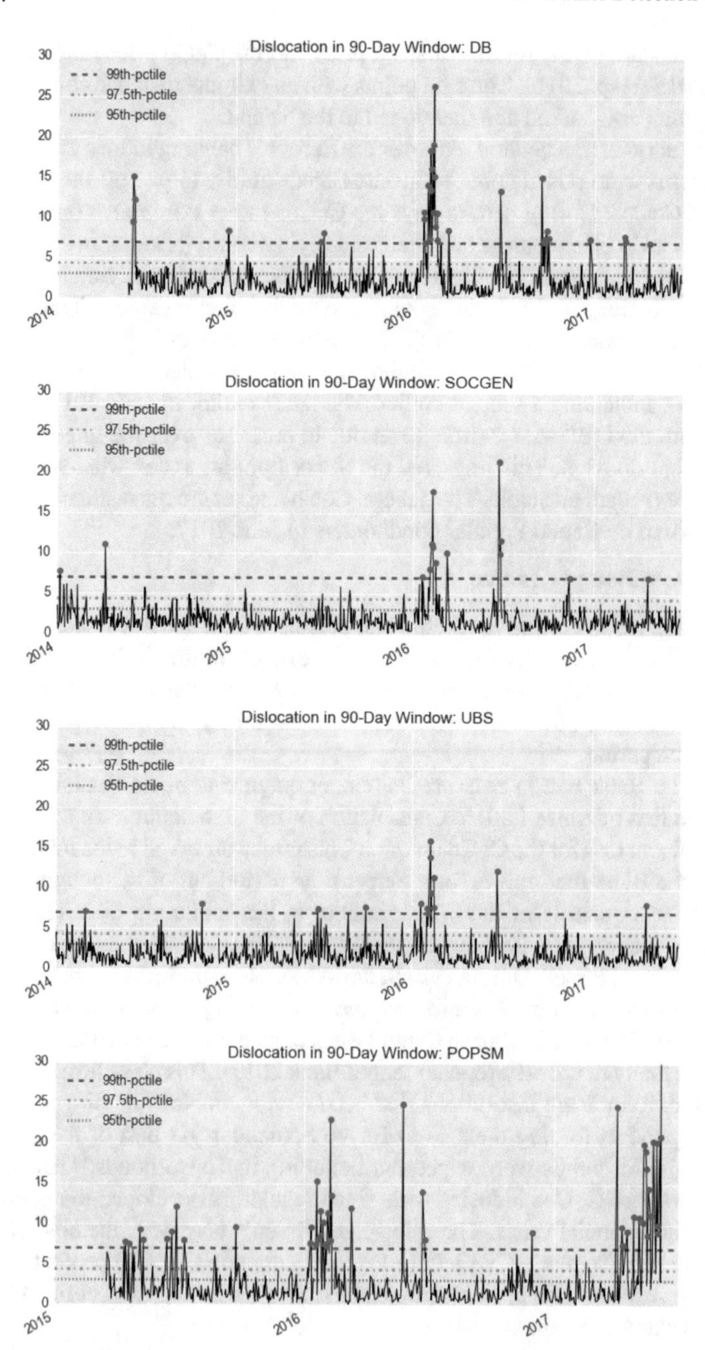

Fig. 6.10 Robust Mahalanobis Distance for CoCos averaged per issuer

Banco Popular shoots again above the indicated boundary. From that point onwards, the distance shoots multiple times above the boundary in all the upcoming weeks. Hence the trigger of Banco Popular CoCo in June 2017 is not unexpected from this data mining exercise.

6.3 Conclusion

The new risk measure, called VEV, entered the PRIIPs guidelines and takes into account the fat-tail and skewed distributions. By application of this measure as described by the regulation, one should be aware of the impact in defining the length of the recommended holding period. In case the period length is set too high, the VEV will be equivalent with volatility. Also the skewness and kurtosis values in the Cornish-Fisher expansion are parameters and do not coincide with the actual kurtosis and skewness of the instrument. Furthermore the formula is only applicable in a certain range of parameter values which is referred to as the domain of validity. The VEV measure denotes however that the CoCos are behaving as described by their hybrid nature between the equity and fixed income asses.

On the financial markets, we observe extreme CoCo price moves together with extreme moves in the underlying equity. This relation is clear by construction of the CoCo asset class. However the detection of outliers in the CoCo market should be performed by taking into account multiple variables like the CoCo market returns and the underlying equity return. Based on a robust multiple-dimension distance we can detect CoCos that are outlying compared to previous time periods but taking into account extreme moves of the market situation as well. We detected with the MCD algorithm as outlier the Deutsche Bank CoCo and the Banco Popular CoCo. Deutsche bank had to reassure its coupon payments of its outstanding CoCos during the first quarter of 2016. On June 7, 2017 the first write-off of CoCos industrywide has occurred for Banco Popular CoCo since initiation of these instruments. Every investor in CoCos should be aware that the high coupon is a compensation for the high risks.

Chapter 7
Conclusion

CoCos are hybrid high-yield instruments that contain an automatically triggered loss-absorption mechanism. These securities convert into equity or experience a write-down when the issuing financial institution is in a life-threatening situation. Hence CoCo bonds automatically improve the solvency of the issuing financial institution in times when it would otherwise have difficulties to raise capital levels. Furthermore conversion CoCos automatically increase the equity basis in times of stress.

Contingent convertible bonds are created to provide a cushion for the issuing bank in times of stress and reduce the cost of governmental bail-out with taxpayers' money. This allows CoCos to count as regulatory capital. In a Basel III setting, the CoCo bonds can count as Tier 2 up to 2% of the RWA or as Additional Tier 1 regulatory capital up to 1.5% of the RWA. AT1 CoCo bonds are more stringent bonds given the fact that they are perpetual and their first call date has to be at least 5 years after the issue date of the bond. Furthermore a particular property of the coupons distributed by such an AT1 CoCo bond can be cancelled. Such a cancellation would not be considered as a default, in contrast with the cancellation of coupon payments on T2 bonds or senior bonds.

CoCo bonds are relatively new and interesting instruments with a high fixed coupon level typically around 6–7% of the notional value but they bear a lot of risk. For a full write-down CoCo the loss is 100% for the investor whereas a conversion CoCo can result in shares with a total recovery rate of 10–30% depending on the type of CoCo. This payout is as such very digital due to the high probability of a high coupon over a long term, and a small probability of an extreme loss in a stress situation. Purchasers of CoCos have included retail investors, hedge funds, asset managers and private banks. CRD IV awares for the fact that no capital guarantee is included. It is also not possible to enjoy enhancement of seniority for CoCo bond holders. The CoCo instruments contain many other risks such as the coupon cancellation, the extension risk and liquidity risk and are due to their complexity excluded for retail selling in the UK.

Modeling CoCos is not straightforward since multiple risk components are not easy to quantify. In this book three different market implied pricing models are discussed. These models create the possibility to grasp different insights in CoCos,

J. De Spiegeleer et al., *The Risk Management of Contingent Convertible (CoCo) Bonds*, SpringerBriefs in Finance, https://doi.org/10.1007/978-3-030-01824-5_7

but also in general to the safeness of the current banking sector based on insights in the CET1 ratio. We introduced the pricing models in a Black–Scholes stock price setting. Although the model has its well-known disadvantages of a constant volatility parameter and underestimation of the tail risk, we were able to describe the price of a CoCo in a tracktable way resulting in different outcomes. For example, the difficulty in these financial products lies in their different characteristics which are hard to compare like the trigger type, conversion type, maturity, coupon cancellation etc. However, the implied barrier methodology from these pricing models can be used as a tool to compare CoCos with different characteristics. As a conclusion, all pricing models have assumptions and simplifications compared to the real market price but their usefulness is in creating insights which makes them interesting to investigate.

The sensitivity analysis of the CoCo price resulted in first place to estimates for the Greeks. The hedging strategy corresponding with these Greeks implies investments in the underlying equity and credit market of the specific CoCo issuer. This requires different investments for each different CoCo issuer in the portfolio. The sensitivity with respect to specific underlying markets can be translated towards a more general sensitivity with respect to some overall market indices with the beta coefficients. The hedging strategy taken into account the beta values will reduce the costs and decrease the follow-up efforts of the hedging strategy for a CoCo portfolio. The fitting of a simple regression model to the market CoCo price revealed the credit spread as a significant variable in modeling CoCo price moves. Hence these regression models indicate the importance of the debt side of a CoCo. Whereas the EDA pricing formula used in the derivation of the Greeks, looks at the CoCo price from an equity perspective. Based on the case study results, we can point out the importance of the credit risk in non-stress situations and the equity risk in a stress situation. Furthermore the VEV risk measure denoted the CoCos are behaving as described between the equity and fixed income asses.

We started the investigation of the derivatives CoCo price models in a Black–Scholes context. However, this stock price model has significant drawbacks such as a constant volatility and the underestimation of tail risk for the underlying stock. In reality, volatility changes with the strike price and the maturity, resulting in the so called volatility-smile. To see the impact of the volatility-smile on the CoCo prices, we put the Heston stock price model at work as a more adequate alternative to the Black–Scholes model. As such we can investigate the impact of skew on the pricing of CoCo bonds by employing a stochastic volatility model (Heston) able of capturing the market skew accurately. We operate in a market implied setting and use the EDA and CDA derivatives approach for the pricing of the CoCo bond. We observe a material impact on the price of CoCos up to 10% due to different skew. CoCos are hence significantly skew sensitive and advanced models are appropriate to accurately capture related risks in the assessment of CoCos.

Contingent convertible bonds can also be seen as derivative instruments contingent on the CET1 level. In this perspective, a CoCo market price is just the price of a derivative and hence is containing forward looking information or at least the market's anticipated view on the financial health of the institution and the level of the relevant trigger. The unadjusted distance to the trigger, i.e. the distance between the current

CET1 ratio and its trigger level, is a weak measure to quantify the embedded risk of a contingent convertible. The CET1 level is a static picture and does not inform us a lot about the business risk of a particular financial institution. The notion of implied CET1 volatility is introduced and used to define a risk-adjusted distance to trigger. The CET1 volatility adjusted distance to trigger has much more explanatory power in describing CoCo spreads than pure distance to trigger.

Different CoCos issued by the same bank and sharing a similar CET1 ratio have almost identical implied CET1 volatility levels. The same results confirm the difference in market risk between Tier 2 and Additional Tier 1 CoCo bonds. The ability to obtain an implied level for the CET1 volatility offers furthermore an other interesting result. The implied coupon cancellation level can be estimated.

Any anomaly on the financial markets might need extra attention from regulatory perspective or cause trading opportunities. On the financial markets, we observe extreme CoCo price moves together with extreme moves in the underlying equity. This relation is clear by construction of the CoCo asset class. However the detection of outliers in the CoCo market should be performed by taking into account multiple variables like the CoCo market returns and the underlying equity return. Based on a robust multiple-dimension distance we can detect CoCos that are outlying compared to previous time periods but taking into account extreme moves of the market situation as well. The developed data mining technique is incorporating a forward looking view by comparing historical data with current CoCo market prices. With the MCD algorithm it was possible to detect as outlier the Deutsche Bank CoCo and the Banco Popular CoCo. Deutsche bank had to reassure its coupon payments of its outstanding CoCos during the first quarter of 2016. On June 7, 2017 the first write-off of CoCos industry-wide has occurred since initiation of these instruments. This was part of the resolution rules for the Spanish bank Banco Popular.

References

Albrecher, H., P. Mayer, W. Schoutens, and J. Tistaert (2007). "The Little Heston Trap". In: Wilmott Magazine, 83–92.

Albul, B., D.M. Jaffee, and A. Tchistyi (2013). "Contingent Convertible Bonds and Capital Structure Decisions".

Allen, H. J. (2012). CoCos can drive markets cuckoo. Lewis & Clark Law Review.

Allen, L. and Y. Tang (2016). "What's the contingency? A proposal for bank contingent capital triggered by systemic risk". In: Journal of Financial Stability 26, pp. 1–14. http://www.sciencedirect.com/science/article/pii/S157230891630047X.

Altman, E. I. and V. M. Kishore (1996). "Almost everything you wanted to know about recoveries on defaulted bonds". In: Financial Analysts Journal 52.6, pp. 57–64.

Avdjiev, S., B. Bogdanova, and A. Kartasheva (2013). "BIS Quarterly Review". In: International banking and financial market developments. Ed. by C. Borio, D. Domanski, C. Upper, S. Cecchetti, and P. Turner. Chap. CoCos: a primer.

Basel Committee on Banking Supervision (2010a). "Basel III: A global regulatory framework for more resilient banks and banking systems". In: Consultative Document.

Basel Committee on Banking Supervision (2010b). Basel III: A global regulatory framework for more resilient banks and banking systems. Tech. rep. Bank for International Settlements.

Basel Committee on Banking Supervision (2010c). "Proposal to ensure the loss absorbency of regulatory capital at the point of non-viability". In: Consultative Document.

Basel Committee on Banking Supervision (2013). "A brief history of the Basel Committee". In: Consultative Document.

Basel Committee on Banking Supervision (2014). "Basel III leverage ratio framework and disclosure requirements". In: Consultative Document.

Benczur, P., G. Cannas, J. Cariboni, F. Di Girolamo, S. Maccaferri, and M. P. Giudici (2016). "Evaluating the effectiveness of the new EU bank regulatory framework: A farewell to bail-out?" In: Journal of Financial Stability.

Brigo, D., J. Garcia, and N. Pede (2015). "CoCo Bonds Pricing with Credit and Equity Calibrated First-Passage Firm Value Models". In: International Journal of Theoretical and Applied Finance 18.3, p. 31.

Brinner, M. (1974). "On the Stability of the Distribution of the Market Component in Stock Price Changes". In: Journal of Financial and Quantitative Analysis 9, pp. 945–961.

Calomiris, C. and R. J. Herring (2013). "How to Design a Contingent Convertible Debt Requirement That Helps Solve Our Too-Big-to-Fail Problem". In: Journal of Applied Corporate Finance 25.2, pp. 21–44.

J. De Spiegeleer et al., *The Risk Management of Contingent Convertible (CoCo) Bonds*, SpringerBriefs in Finance, https://doi.org/10.1007/978-3-030-01824-5

Calomiris, C. and R. Herring (2011). "Why and how to desing a contingent convertible debt requirement". In: Rocky times: new perspectives on financial stability. Brookings/NICMR Press. Chap. 5.

Campolongo, F., J. De Spiegeleer, F. Di Girolamo, and W. Schoutens (2017). "Contingent Conversion Convertible Bond: New avenue to raise bank capital." In: International Journal of Financial Engineering 4.

Carr, P. and D. Madan (1999). "Option valuation using the fast fourier transform". In: Journal of Computational Finance 2, pp. 61–73.

Cerioli, A. (2010). "Multivariate outlier detection with high-breakdown estimators." In: Journal of the American Statistical Association 105.489, pp. 147–156.

Cheridito, P. and X. Zhikai (2013). "A reduced form CoCo model with deterministic conversion intensity". Available on SSRN: http://ssrn.com/abstract=2254403.

Chung, T. K. and Y. K. Kwok (2016). "Enhanced Equity-Credit Modeling for Contingent Convertibles". In: Quantitative Finance 16.10, pp. 1–17.

Commission of Experts (2010). Final report of the Commission of Experts for limiting the economic risks posed by large companies. Tech. rep. Swiss National Bank.

Corcuera, J. M., J. De Spiegeleer, A. Ferreiro-Castilla, A. E. Kyprianou, D. B. Madan, and W. Schoutens (2013). "Pricing of Contingent Convertibles under Smile Conform Models". In: Journal of Credit Risk 9.3, pp. 121–140.

Corcuera, J. M., J. De Spiegeleer, J. Fajardo, H. Jönsson, W. Schoutens, and A. Valdivia (2014). "Close form pricing formulas for CoCa CoCos". In: Journal of Banking and Finance 1.

Cornish, E. A. and R. A. Fisher (1938). "Moments and Cumulants in the Specification of Distributions". In: Review of the International Statistical Institute 5.4, pp. 307–320.

D' Souza, A., B. Foran, G. E. Hafez, C. Himmelberg, Q. Mai, J. Mannoia, R. Ramsden, and S. Romanoff (2009). Ending 'Too Big To Fail'. Tech. rep. New York: Goldman Sachs Global Markets Institue.

De Spiegeleer, J., M. Forys, and W. Schoutens (2012). "Euromoney Encyclopedia of Debt Finance". In: Euromoney Books. Chap. Contingent convertibles: introduction to a new asset class, pp. 113–126.

De Spiegeleer, J., I. Marquet, and W. Schoutens (2017). "Data Mining of Contingent Convertible Bonds". In: Journal of Financial Management, Markets and Institutions 2, pp. 147–168. https://doi.org/10.12831/88825.

De Spiegeleer, J. and W. Schoutens (2010). The Handbook of Convertible Bonds: Pricing, Strategies and Risk Management. Ed. by Wiley. Wiley.

De Spiegeleer, J. and W. Schoutens (2011). Contingent Convertible CoCo-Notes: Structuring & Pricing. London: Euromoney Institutional Invester PLC.

De Spiegeleer, J. and W. Schoutens (2012a). "Pricing Contingent Convertibles: A Derivatives Approach". In: Journal of Derivatives 20.2, pp. 27–36.

De Spiegeleer, J. and W. Schoutens (2012b). "Steering a bank around a death spiral: Multiple Trigger CoCos". In: Wilmott Magazine 2012.59, pp. 62–69.

De Spiegeleer, J. and W. Schoutens (2013). "Multiple Trigger CoCos: Contingent debt without death spiral risk". In: Financial Markets, Institutions and Instruments 22.2.

De Spiegeleer, J. and W. Schoutens (2014). "CoCo Bonds With Extension Risk". In: Wilmott Magazine 2014.71, pp. 78–91.

De Spiegeleer, J., W. Schoutens, and C. Van Hulle (2014). The Handbook of Hybrid Securities: Convertible Bonds, Coco Bonds and Bail-in. Ed. by Wiley.Wiley.

De Spiegeleer, J., S. Höcht, I. Marquet, and W. Schoutens (2017a). "CoCo bonds and implied CET1 volatility". In: Quantitative Finance 17.6, pp. 813–824. https://doi.org/10.1080/14697688.2016.1249019.

De Spiegeleer, J., M. Forys, I. Marquet, and W. Schoutens (2017b). "The impact of skew on the pricing of CoCo bonds". In: International Journal of Financial Engineering 04.01. https://doi.org/10.1142/S2424786317500128.

ECB-public (2016). SSM SREP Methodology Booklet. Tech. rep. Supervisory Review and Evaluation Process. European Central Bank.
European Central Bank (2017). ECB determined Banco Popular Espanol S.A. was failing or likely to fail. Press Release.
European Commission (2017). Annexes to the Commission Delegated Regulation (EU). Tech. rep. Official Journal of the European Union.
FCA (2014). Temporary product intervention rules: Restrictions in relation to the retail distribution of contingent convertible instruments. Tech. rep. Financial Conduct Authority.
FCA (2015). Restrictions on the retail distribution of regulatory capital instruments: Feedback to CP14/23 and final rules. Tech. rep. Financial Conduct Authority.
Fiamma, S., J. Heinberg, and D. Lewis (2012). Tax Treatment of Additional Tier 1 Capital under Basel III. Tech. rep. Allen & Overy LLP.
Fitch Solutions (2011). Pricing and calibration of contingent capital with a structural approach. Tech. rep. Fitch Group.
Flannery, M. J. (2009). Stabilizing Large Financial Institutions with Contingent Capital Certificates. Tech. rep. Graduate School of Business Administration, University of Florida.
Green, C. G. and R. D. Martin (2014). "An extension of a method of Hardin and Rocke, with an application to multivariate outlier detection via the IRMCD method of Cerioli." Working Paper.
Guarascio, F. (2018). EU warned of wind-down risk for Spain's Banco Popular - source. Available at Reuters: http://uk.reuters.com/article.
Hajiloizou, C., Y. di Mambro, and K. Gheeraert (2015). European Banks - The CoCo Handbook Vol. 6. Tech. rep. Barclays Capital - Credit Research.
Hajiloizou, C., Y. di Mambro, D. Winnicki, J. Simmonds, and S. Gupta (2014). European Banks - The CoCo Handbook Vol. 5. Tech. rep. Barclays Capital -Credit Research.
Haldane, A. G. (2011). Capital Discipline. speech given at the American Economic Association, Denver.
Hardin, J. and D. M. Rocke (2005). "The distribution of robust distances". In: Journal of Computational and Graphical Statistics 14.4, pp. 928–946.
Hendrickx, N. (2016–2017). "Liquidity of CoCo bonds: a data analytical approach". MA thesis. KU Leuven.
Heston, S. (1993). "A closed-form solutions for options with stochastic volatility with applications to bond and currency option." In: Review of Financial Studies 6.59, pp. 327–343.
Hickman, W. B. (1958). Corporate bond quality and investor experience. Tech. rep. Princeton University Press.
Hilscher, J. and A. Raviv (2014). "Bank stability and market discipline: The effect of contingent capital on risk taking and default probability". In: Journal of Corporate Finance 29, pp. 542–560.
Hoyle, B., M. M. Raul, K. Paech, C. Bonnett, S. Seitzl, and J. Weller (2016). "Anomaly detection for machine learning redshifts applied to SDSS galaxies". In: Monthly Notices of the Royal Astronomical Society.
Hubert, M. and M. Debruyne (2009). "Minimum covariance determinant". In: Wiley Interdisciplinary Reviews: Computational Statistics 2, pp. 36–43.
Hull, J. C. (2010). "Risk Management and Financial Institutions". In: Pearson. Chap. 11, pp. 221–234.
JP Morgan (1999). "Pricing Exotics under the Smile". In: Risk Magazine, pp. 72–75.
Jaworski, P., K. Liberadzki, and M. Liberadzki (2017). "How does issuing contingent convertible bonds improve bank's solvency? A Value-at-Risk and Expected Shortfall Approach". In: jounal of economic modelling 60, pp. 162–168.
Leoni, P. (2014). The Greeks and Hedging Explained. Palgrave Macmillan UK.
Liberadzki, K. and M. Liberadzki (2016). Hybrid Securities: Structuring, Pricing and Risk Assessment. Palgrave Macmillan.
Lintner, J. V. (1965). "Security prices, risk and maximal gains from diversification". In: Journal of Finance 20.4, pp. 587–615.

Madan, D. B. and W. Schoutens (2011). "Conic coconuts: the pricing of contingent capital notes using conic finance". In: Mathematics and Financial Economics 4.2, pp. 87–106.
Maes, S. and W. Schoutens (2012). "Contingent Capital: An in-depth Discussion". In: Economic Notes 41.1/22, pp. 59–79.
Maillard, D. (2012). A User's Guide to the Cornish Fisher Expansion. Tech. rep. Available at SSRN: https://ssrn.com/abstract=1997178. Conservatoire National des Arts et Métiers (CNAM); Amundi Asset Management.
Mandelbrot, B. (1963). "The variation of certain speculative prices". In: The Journal of Business 4, 394–419.
Martynova, N. and E. Perotti (2015). "Convertible bonds and bank risk-taking". In: DNB Working Paper 480. De Nederlandsche Bank, pp. 1–42.
McDonald, R. L. (2013). "Contingent capital with a dual price trigger". In: Journal of Financial Stability 9.2, pp. 230–241.
Mehta, N. (2016). Cocos remain liquid despite recent volatility. IHSMarkit.com. Markit Commentary Credit.
Merrouche, O. and M. Mariathasan (2014). "The manipulation of basel riskweights". In: Journal of Financial Intermediation 23.3, pp. 300–321.
Nelder, J.A. and R. Mead (1965). "A simplex method for function minimization". In: The Computer Journal 7.4, pp. 308–313.
Nordal, K. B. and N. Stefano (2014). Contingent Convertible Bonds (CoCos) issued by European banks. Tech. rep. 19. Staff memo. Norges Bank.
Pennacchi, G. (2010). "A Structural Model of Contingent Bank Capital". Available at SSRN: https://ssrn.com/abstract=1595080.
Pennacchi, G., T. Vermaelen, and C. Wolff (2014). "Contingent capital: The case for COERCs". In: Journal of Financial and Quantitative Analysis 49.3, pp. 541–574.
Pison, G., S. Van Aelst, and G. Willems (2002). "Small sample corrections for LTS and MCD". In: Metrika 55, pp. 111–123.
Rousseeuw, P. J. and K. van Driessen (1999). "A Fast Algorithm for the Minimum Covariance Determinant Estimator". In: Technometrics 41.3, pp. 212–223.
Rousseeuw, P. J. and B. C. van Zomeren (1990). "Unmasking multivariate outliers and leverage points". In: Journal of American Statistics Association 58, pp. 633–651.
Rousseeuw, P. J., M. Debruyne, Engelen S., and Hubert M. (2006). "Robustness and Outlier Detection in Chemometrics". In: Critical Reviews in Analytical Chemistry 36, pp. 221–242.
Rubinstein, M. and E. Reiner (1991). "Unscrambling the Binary Code". In: Risk Magazine 4, pp. 75–83.
Schoutens, W. (2008). The world of Variance Gamma. Tech. rep. KU Leuven.
Schoutens, W., J. Tistaert, and E. Simons (2004). "A Perfect Calibration ! Now What ?" In: Wilmott Magazine 2.
Sharpe, W. F. (1964). "Capital Asset Prices: A Theory of Market Equilibrium Under Conditions of Risk". In: Journal of Finance 19, pp. 425–442.
Su, L. and M. O. Rieger (2009). How likely is it to Hit a Barrier? Theoretical and emperical Estimates. Tech. rep. Working Paper No. 594. National Centre of Competence in Research, Financial Valuation and Risk Management.
Thompson, C., D. Schäfer, and B. McLannahan (2014). "Deutsche Bank leads wave of 'coco' issuance". In: Financial Times.
Weber A. an Glover, J., B. Groendahl, and S. Sirletti (2017). CoCo Investors May Lose Payout Priority as EU Revamps Laws. Bloomberg Markets.

Printed in the United States
By Bookmasters